AF412481

The NanoPositioning Book

Thomas R Hicks

Paul D Atherton

with contributions from

Ying Xu

Malachy McConnell

Queensgate Instruments Ltd

Queensgate House

Waterside Park

Bracknell

Berkshire RG12 1RB

Published by Queensgate Instruments Limited
Waterside Park, Bracknell, Berkshire, RG12 1RB United Kingdom.

Tel: 01344 484111
Fax: 01344 484115
E-mail nano@queensgate.co.uk
Web site http:www.queensgate.com

First Published in the United Kingdom 1997

British Library Cataloguing in Publication Data
A CIP catalogue record for this book can be obtained from the
British Library.

ISBN 0 9530658 0 4

Produced in the United Kingdom by
Product Promotion Group Limited, Sunbury on Thames TW16 7HB
and bound by Biddles Limited, Guildford and King's Lynn.

CONTENTS

Contents

Contents

When it comes to specifying performance parameters of positioning and measuring devices for moving and measuring to nanometre precision, one immediately comes across the problem of defining what one actually means by various terms in common usage. What is accuracy? Is it the same as precision? Is non-linearity the same as linearity? What reference plane do you take to define roll, pitch and yaw? Of course many international standards have been written to help sort this out, and indeed they do help. But there are contradictions and ambiguities. Some standards warmly embrace the concept of a 'true' value, others abhor the idea. Some talk of bias, others trueness and systematic error. What terms should one adopt? In an attempt to answer these questions and standardise terms within Queensgate, the NanoPositioning Book has evolved. It started as a need for a list of definitions and ended as this book, via a lot of head scratching, standards reading, free and frank discussion and downright imposition!

This book is not intended to replace international standards. Rather it draws on them to define consistent terms for use in specifying Nanometre Precision Mechanisms. On the way it gives an insight into the technologies involved in Queensgate nanopositioning mechanisms. Chapter 1 gives a brief introduction to the field; Chapter 2 gets down to the nitty-gritty of measurement definitions (this one's a bit dense on first reading: skip over it if you like); Chapter 3 gives some background to the servo control systems used in our products and Chapter 4 the materials. Chapters 5 and 6 deal with the core technologies of measuring and moving small distances and then Chapter 7 brings these together to describe some practical mechanisms and how their performance is measured and specified. The Glossary gives concise definitions of some of the terms used, in name order and symbol order, and I hope the index will help in finding specific topics.

This book has been prepared using Microsoft Word for Windows 95 Version 7.0a; Excel for Windows 95 Version 7.0a; the Math Works Inc. Matlab version 4.0 and Simulink. Historic pictures where found on the web at www.arttoday.com and are reproduced within the terms of their license agreement.

Definitions have been taken where possible from International and National Standards, with International ones taking precedence when necessary. Relevant ones are:

- ISO 5725-1,-2,-3,-4,-6:1994(E), Accuracy (trueness and precision) of measurement methods and results.

- PD 6461 Part 2 1980; Part 3 1995, Vocabulary of Metrology (BSI)

- BS5233:1986, Terms used in metrology.

- BS 6808 Part 1, 1987 Coordinate measuring machines.

- ISO 31-0,-1,-3,-5,-11:1993, Quantities, units and symbols

Many people have helped with this book: in particular I would like to thank Dr. Mike Downs of the UK National Physical Laboratory; Dr. Derek Chetwynd of the Centre for Nanotechnology and Microengineering, University of Warwick; the other directors and staff of Queensgate in the UK, USA and Japan, especially (in alphabetical order) Hitoshi Ariu, Colin Chambers, Warren Gutheil, Graham Jones, Liz Kirk, Jayesh Patel, Phil Rhead, Jerry Russell, Sam Salloum, Sean Staines, for input and helpful comments. And also of course our customers, who have to live with the consequences!

TRH April 1997.
trh@queensgate.co.uk

Please visit:
www.queensgate.com

What is it all about?

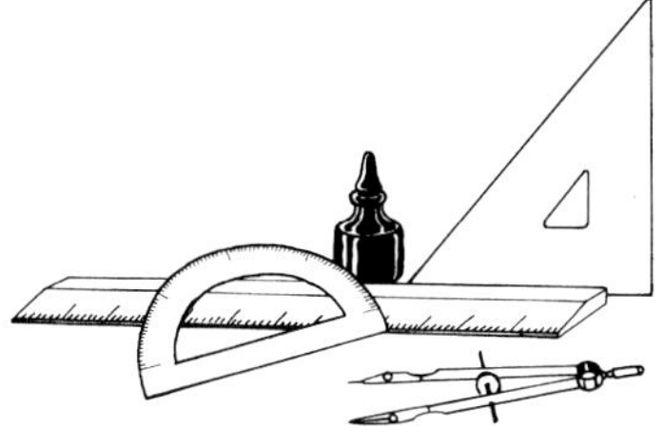

This book describes and explains the technology used in Queensgate NanoPositioning products. It is designed to provide sufficient background information for an engineer to be able to understand how to interpret the technical data provided with Queensgate products, and to design systems which use Queensgate products in an optimal way. A great deal of this information has been developed by Queensgate engineers during the past 25 years. Some of it comes from obscure library books, and some of it is not available anywhere else. A list of references at the end of each chapter will help those interested to learn more. A glossary is also provided to define carefully a number of much abused terms such as accuracy, precision, resolution. Alternatively you can ask us.... NanoPositioning from the Experts....

In this section:

NanoPositioning, the technology of moving and measuring with sub-nanometre precision, helps researchers and manufacturers in the production of some of the most advanced products in use today. For example, wafer steppers use this technology to make silicon chips with line widths down to 200 nm; Scanning Probe Microscopes are used to establish how well such chips are made; and the introduction of M-R head technology allows 5 Gigabyte disks to become the norm (1997), while 23 Gigabyte disk stacks are now available for desk top PCs. These machines, and the machines which make these machines, combine advanced optical design with advanced motion control technology which can position components to accuracies of a nanometre or below. Piezoelectric devices have the potential to meet the resolution requirements. However, because piezoelectric devices are non-linear and exhibit hysteresis, they require an external sensor to control their position. The capacitance micrometer is ideally suited to this task, being small, simple, and with sub-atomic intrinsic resolution capability. To make a NanoMechanism involves combining the piezoelectric actuator and capacitance position sensor with flexure hinges and mechanical amplifiers in such a way as to produce a device

---

<table>
<tr><td colspan="3" align="center">Units and Prefixes.</td></tr>
<tr><td colspan="3">The prefixes nano and micro are used throughout this book to indicate multiples or sub-multiples of a quantity. The full list with symbols is:</td></tr>
<tr><td>10^{-24}</td><td>yocto</td><td>y</td></tr>
<tr><td>10^{-21}</td><td>zepto</td><td>z</td></tr>
<tr><td>10^{-18}</td><td>atto</td><td>a</td></tr>
<tr><td>10^{-15}</td><td>femto</td><td>f</td></tr>
<tr><td>10^{-12}</td><td>pico</td><td>p</td></tr>
<tr><td>10^{-9}</td><td>nano</td><td>n</td></tr>
<tr><td>10^{-6}</td><td>micro</td><td>μ</td></tr>
<tr><td>10^{-3}</td><td>milli</td><td>m</td></tr>
<tr><td>10^{3}</td><td>kilo</td><td>k</td></tr>
<tr><td>10^{6}</td><td>mega</td><td>M</td></tr>
<tr><td>10^{9}</td><td>giga</td><td>G</td></tr>
<tr><td>10^{12}</td><td>tera</td><td>T</td></tr>
<tr><td>10^{15}</td><td>peta</td><td>P</td></tr>
<tr><td>10^{18}</td><td>exa</td><td>E</td></tr>
<tr><td>10^{21}</td><td>zetta</td><td>Z</td></tr>
<tr><td>10^{24}</td><td>yotta</td><td>Y</td></tr>
</table>

Thus a nanometre (nm) is 10^{-9}m. To put things in perspective the diameter of the hydrogen atom is about 100pm; an electron is about 6fm across; the average beard grows about 10nm in a second and an attoparsec is about a tenth of a foot.

capable of making pure orthogonal motions with sub nanometre precision.

In this book we describe the basic theory, the capabilities, the limitations and the engineering background information behind capacitance position sensing, piezoelectric translators and NanoMechanisms.

ELECTRO-ACTIVE MATERIALS.

Moving tiny distances

Certain materials exhibit properties that cause them to undergo dimensional changes in an electric field. These properties are commonly known as the piezoelectric and electrostrictive effects. Typically these effects are very small in natural materials. For example quartz exhibits a piezoelectric effect corresponding to a maximum strain (extension per unit length) of about 1 part per million. These effects can be optimised by appropriate doping of ceramics and are tailored to a variety of applications, such as loudspeakers and ultra-sonic cleaners. The piezoelectric effect is reversible so that an applied stress will produce a voltage. This property is commonly used in gas igniters and microphones. The familiar acronym PZT comes from lead (Pb) zirconate (ZrO_3) titanate (TiO_3), a complex ceramic designed especially for applications requiring electro-mechanical conversion of some kind. Similarly the acronym for electrostrictive materials is PMN [lead (Pb) magnesium (Mg) niobate (NiO_3)]. For a detailed discussion of these materials see a text such as that by Burfoot and Taylor, 1979.

In NanoPositioning applications one great advantage of such materials is the combination of sub-atomic resolution (picometre or below) with very high mechanical stiffness. This advantage outweighs their primary disadvantage which is their extremely limited range. The maximum strain achievable is typically 0.1% for reliable operation, about 1000 times greater than quartz. Thus a device with 100 μm range would have to be 100 mm long. To reduce the operating voltage to a low level, these devices are manufactured in stacks of very thin layers. For example a 20 mm long stack might have 200 layers each 100 μm thick, and expand by 15 μm for 100 V. Typically these layers are produced using the tape-casting techniques developed for the manufacture of capacitors. Such a stack is typically very strong in compression and able to generate a force of about 750 N. This high stiffness results in very high resonant frequencies, enabling devices to move at high speed in a controlled manner. Because they can generate high forces it is also possible to amplify their motion mechanically at the expense of lower frequency operation.

CAPACITANCE MICROMETRY

Measuring tiny distances

Capacitance micrometry is a very sensitive technique for detecting small displacements. It works by detecting the change in impedance of a parallel plate capacitor as the spacing or area changes. Displacements as small as 10^{-14} m, about the diameter of an electron, or 10 000 times smaller than an atom, have been measured using this technique.

A capacitance micrometer is essentially quite simple to make. Two conducting electrodes, often metallic films or shims, are separated by about 500 µm, and have dielectric between them, usually air or a vacuum. With a 12 mm diameter pad a capacitance of about 2pF is achieved. This capacitance is compared to some kind of reference using an ac bridge. Any change in the micrometer capacitance, due to a change in its area or spacing, is demodulated and presented as a dc signal proportional in some way to the change in capacitance and related to the change in area, spacing or dielectric constant of the micrometer capacitor.

Capacitance micrometers have the advantage of very high position resolution (far in excess of most laser interferometers), zero hysteresis, zero power dissipation at the point of measurement, high linearity (0.01% is possible), insensitivity to crosstalk, simplicity and the ability to be made from very stable materials, such as Invar or Zerodur. Queensgate's NanoSensors are practical examples of capacitance micrometers.

NANOMECHANISMS

Motion and measurement

By using NanoSensors to monitor the movement of mechanisms and translators it is possible to servo-control the position of these devices to sub-nanometre precision. A simple example of this is the Digital Piezo Translator. Here the motion of the PZT is monitored by the capacitance micrometer. Any hysteresis, drift or creep in the length of the PZT is monitored by the sensor. The output of the sensor is then used to control the voltage on the PZT to form a closed loop system. In this manner the DPT achieves sub-nanometre reproducibility and deviation from perfect linearity of 0.05%. More complex mechanisms combine several axes of motion, along with mechanical amplification of the PZT motion. In order to minimise parasitic motions, i.e. motion which is not purely along a single dimension, and to ensure that these motions are orthogonal, flexure mechanisms are used. Optimisation of these designs is quite complex, requiring advanced design tools and extensive prototyping. These NanoMechanisms include NanoSensors on each axis to ensure sub-nanometre precision.

ABOUT THIS BOOK

Before going into the details of control systems, sensors, positioners and mechanisms it is helpful to define some of the terms used to describe 'accuracy'. This is dealt with in Chapter 2. The concept of accuracy however gives rise to a rather complex series of definitions which may at first be a bit off-putting. Don't worry too much about it: a full understanding is not required to understand the succeeding chapters!

Chapters 3 to 6 discuss the techniques and components that make up NanoMechanisms and Chapter 7 brings these together to describe the properties of some practical systems.

Chapter 8 is a glossary of terms which also acts as an index to more complete discussions.

REFERENCES

Burfoot J.C. and Taylor G.W. Polar Dielectrics and their applications., *The MacMillan Press Ltd., 1979.*

CHAPTER 2. ACCURACY, TRUENESS AND PRECISION

and other hard words

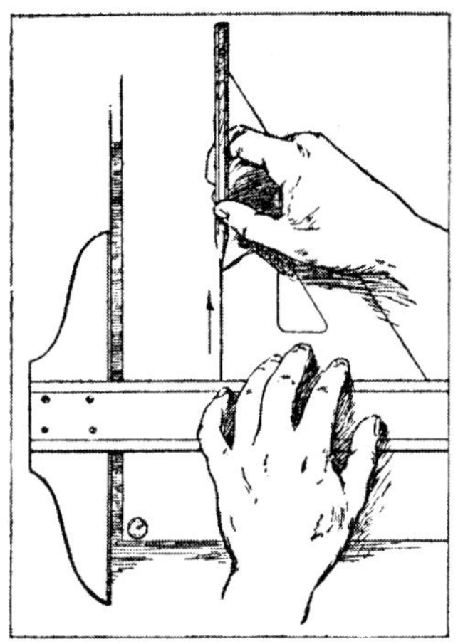

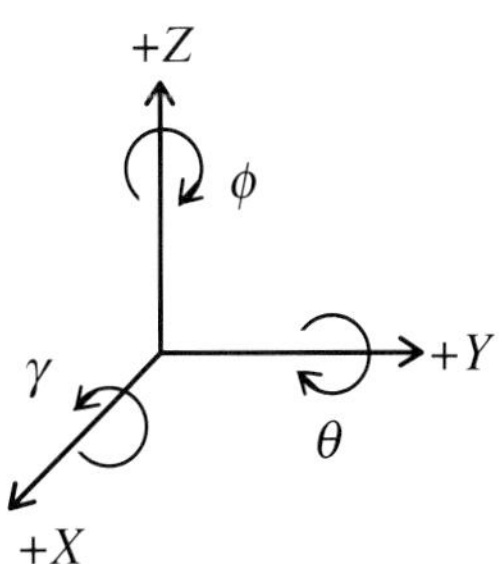

Accuracy, trueness and precision: these words are often used interchangeably but actually mean different things. To ensure uniformity, Queensgate uses definitions based on ISO5725-1:1994(E).

This section is designed to give enough technical background information to allow an engineer or scientist to understand the issues involved in designing and using nanopositioning systems. Here we define various terms required for a full understanding of NanoMechanisms, their specification and the supplied test data.

THE COORDINATE SYSTEM

Everything is relative to something

Fig. 2.1 Coordinate System

First it is necessary to define the coordinates used to describe positions. The obvious system to use for positioning stages is an orthogonal Cartesian Coordinate system. With this one can define a position with its X, Y, Z coordinates and an arbitrary rotation as components of rotation about the X, Y and Z axes. More usefully one can describe a movement or displacement as a change in the X, Y and Z coordinates.

Fig. 2.1 shows the system used. As drawn, the positive X direction is out of the page and the negative X direction down into it. Rotations are described with respect to the X, Y and Z axes in a right-handed sense, thus $+\theta$ is a clockwise rotation when looking along the $+Y$ direction, $+\phi$ is a clockwise rotation when looking along the $+Z$ direction and $+\gamma$ is a clockwise rotation when looking along the $+X$ direction.

Roll, pitch and yaw

Which way is 'up'?

One sees a variety of symbols used to denote rotations about the X, Y and Z axes, for example $\alpha, \beta, \gamma; \theta_1, \theta_2, \theta_3$. We use γ, θ, ϕ to maintain some consistency with the accepted altitude and azimuth symbols θ, ϕ used in spherical polar coordinate systems.

$\delta\gamma, \theta, \phi$ are used to denote roll, pitch and yaw. As these terms depend on the direction of motion, subscripts must be used to fully define them. Thus $\delta_{\theta x}$ is a θ rotation when moving along the +X axis, i.e. pitch. So we have:

roll: $\delta\gamma_x \quad \delta\theta_y \quad \delta\phi_z$
pitch: $\delta\theta_x \quad \delta\phi_y \quad \delta\gamma_z$
yaw: $\delta\phi_x \quad \delta\gamma_y \quad \delta\theta_z$

The sign of the roll, pitch or yaw is obtained by multiplying the sign of the motion direction by the sign of the resultant rotation.

The terms pitch, roll and yaw are often used when talking about rotations. These terms are useful when describing parasitic rotations caused by a linear motion, but **great care** must be taken as they are referred to the *direction of motion* and a concept of *'up'* rather than a defined axis system. For an aeroplane in flight, a rotation about an axis drawn from wing tip to wing tip is pitch; a rotation about an axis drawn down the length of the fuselage is roll and a rotation about a vertical axis is yaw. In the defined Cartesian system if the 'plane is flying along the positive X direction θ is pitch, γ is roll and ϕ is yaw. Using an aeroplane as an example we have arbitrarily assumed that the XY plane is horizontal and Z is up. This *could* also be assumed for motion along Y, but what happens when you fly along Z? Which way is up then? For completeness we must have consistent definitions of the 'horizontal' planes for motion along all the axes. This produces some apparently weird results which highlights the problem of the terms roll, pitch and yaw. We do use the terms however, so to be sure, observe the associated symbols!

Motion along	'Horizontal' plane	'Up'	Roll	Pitch	Yaw
X	XY	+Z	γ	θ	ϕ
Y	YZ	+X	θ	ϕ	γ
Z	ZX	+Y	ϕ	γ	θ

Table 2.1 Roll, pitch and yaw axes

POSITION MEASUREMENT

Where you are and where you think you are

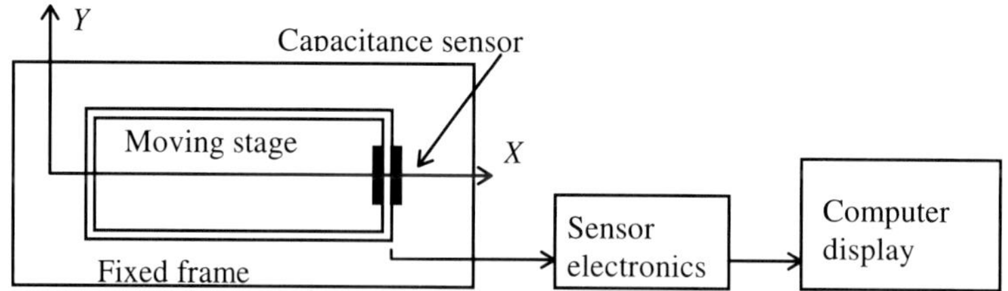

Fig. 2.2 A simple X stage

When talking about position measurement, it is useful to consider a simple one-axis translation stage as shown in Fig. 2.2. The stage is designed to move in the X direction with respect to the fixed frame and a capacitance sensor measures this movement and displays it on a computer screen. We must now introduce the concept of an external, perfect measuring device that can tell us what the 'true' motion or position of the stage is. Let us suppose that the stage moves from a position O, which we will consider as the origin for measurements, to a true position x_p shown in Fig. 2.3. At present we are not concerned with what causes the motion.

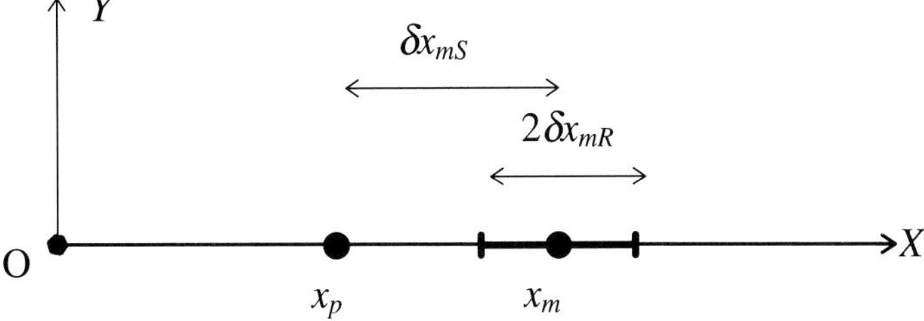

Fig. 2.3. X movement

The sensor will measure the position as x_m which will not be the same as x_p. The difference between the measured position and the true position is the measurement *trueness,* quantified by a systematic error δx_{mS}. Obviously one would like this to be small. The measured position x_m is shown with an error bar as there will be noise and other random errors: every time you make a sample measurement it will be different. The size of the bar defines the measurement *precision,* quantified by a random error δx_{mR}. If δx_{mS} is finite and δx_{mR} is small, one can get precisely the wrong answer!

The general term *accuracy* is a combination of trueness and precision. The overall measurement accuracy is defined by the error

$$\delta x_{mA} = \delta x_{mR} + \delta x_{mS} \qquad (2.1).$$

The rest of this section is devoted to defining the parameters that make up measurement trueness and precision. So what are they? In summary:

Table 2.2 Accuracy and Precision

Property	Constituent parameters	Dealt with in sub-section:
Measurement trueness	Mapping trueness	Measurement Linearity and Mapping
	Abbe error	Cosine and Abbe Errors
	Cosine error	Cosine and Abbe Errors
Measurement precision	Mapping error	Measurement Linearity and Mapping
	Resolution	Measurement Resolution and Noise
	Noise	Measurement Resolution and Noise
	Reproducibility	Measurement Repeatability and Reproducibility
	Repeatability	Measurement Repeatability and Reproducibility

As will be seen, some of the constituent parameters can be broken down further. The full picture is shown in Fig. 2.4:

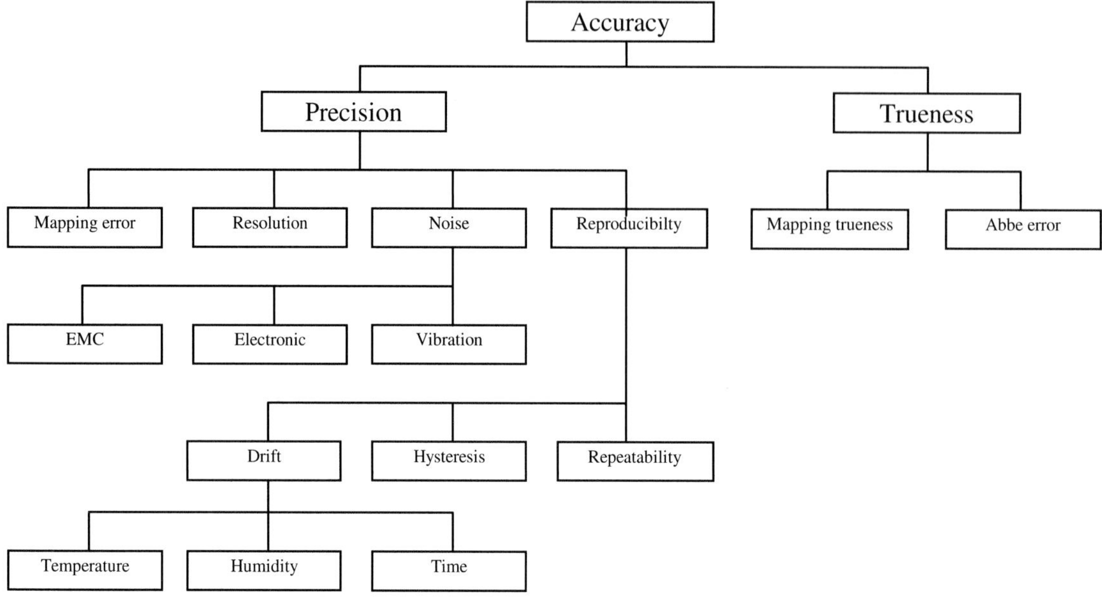

Fig. 2.4.Measurment Accuracy

Measurement Linearity and Mapping

We worry a lot about this

In an ideal world the position displayed on the user's computer would be the same as the true position, that is

$$x_m = x_p \qquad (2.2).$$

The world is almost ideal but not quite, so equation 2.2 must be replaced by some mapping function

$$x_m = f\left(x_p\right) \qquad (2.3).$$

Various forms of measurement mapping function are described in more detail in Chapter 5, Capacitance Sensors, but the simplest is a power series

$$x_m = a_{x0} + a_{x1}x_p + a_{x2}x_p^{\,2} + a_{x3}x_p^{\,3} + a_{x4}x_p^{\,4}\ldots\ldots$$
$$(2.4).$$

Even simpler is a first order power series, or straight line which is equation 2.4 truncated after the term in x_p:

$$x_m = a_{x0} + a_{x1}x_p \qquad\qquad (2.5).$$

Obviously this will not give as good a result as equation 2.4 but it is often adequate. The term a_{x1} is the linear factor describing the relationship between the true stage position as measured by a hypothetical perfectly accurate position sensor and the position measurement passed to the user's computer, that is the *measurement scale factor*. It is the gradient of the best fit straight line (first order power series) to the general mapping function of equation 2.3 over the sensor range.

Mapping Accuracy and Scale Factor Uncertainty

Mapping accuracy is a measure of how well equation 2.4 fits the sensor performance (mapping trueness) and how well the *'a'* coefficients are determined by the manufacturer (mapping precision). Ideally the coefficients will be traceable back to measurements made against international length standards. This is done using a laser interferometer as described in the Appendix to Chapter 7, Nanometre Precision Mechanisms. Briefly a set of 'true' displacements x_p as defined by the interferometer are applied to the sensor and x_m measured. A power series of chosen order is then fitted using linear regression and the *'a'* coefficients with their uncertainties derived.

The *mapping accuracy* is the set of errors on the individual *'a'* coefficients, but normally only a first order power series is used so only a_1 is considered. The mapping accuracy is then the sensor *scale factor uncertainty*, $\delta a_{x1,}$ so

$$a_{x1} = 1 \pm \delta a_{x1} \qquad\qquad (2.6).$$

The scale factor trueness also depends on how well the capacitance sensor is aligned during assembly. For sensors supplied built into a stage this is not an issue for the user, but it will be if the sensors are used stand-alone. This is dealt with in detail in Chapter 5, Capacitance Sensors. The scale factor error is the largest contributor to the measurement trueness, δx_{mS}.

Mapping Error and Linearity

Mapping error is defined with respect to a residual curve obtained by subtracting the fit to the data set (equation 2.4) from the data set. Since the linear regression used to do the fitting minimises the sum of the squares of the deviations from the line, there is usually a bigger residual on one side of the line than the other. We thus define the mapping error as half of the peak to peak deviation compared to the full range, expressed as a percentage.

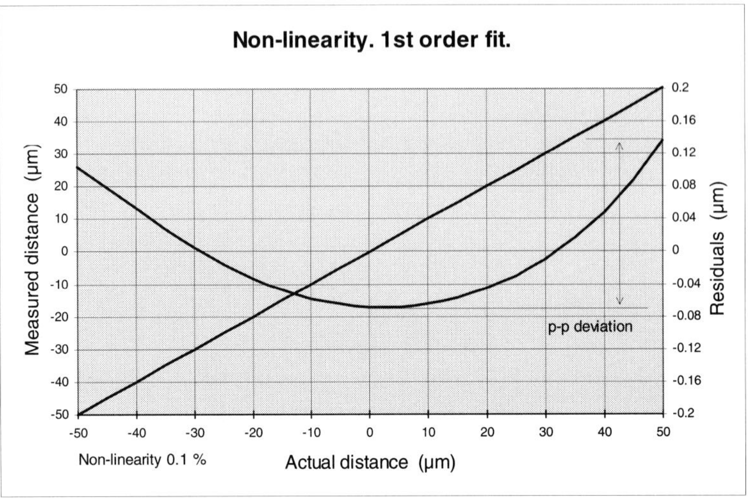

Fig. 2.5. Linearity error

When the mapping function is first order (a straight line), then the mapping error becomes the *linearity error* or *non-linearity*, $\delta x_{m.\text{lin}}$. As an example, a linearity error of 0.1% in a 100 µm range device results in a 100 nm absolute position uncertainty between the 0 µm position and the 100 µm position. This is illustrated in Fig. 2.5. Typically linearity errors of 0.1 % are easily achieved. Below 0.01 % the measurements are limited by the intrinsic linearity of the calibration systems.

Higher order mapping functions

Usually the deviation from linearity is roughly parabolic and in some systems this is easy to compensate for electronically. The result of compensating one, slightly imperfect, parabola with another is usually an 'S' curve of much lower amplitude so the mapping error is much lower. This is equivalent to using the a_{x1} and a_{x2} terms of equation 2.4,

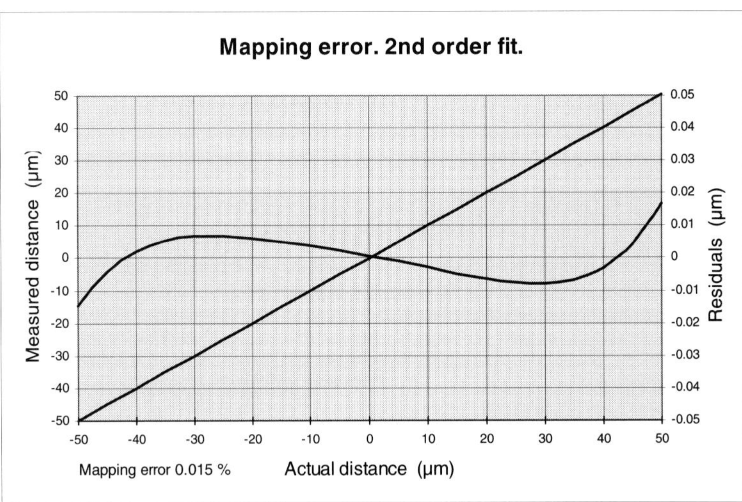

Fig. 2.6. 2nd order fit

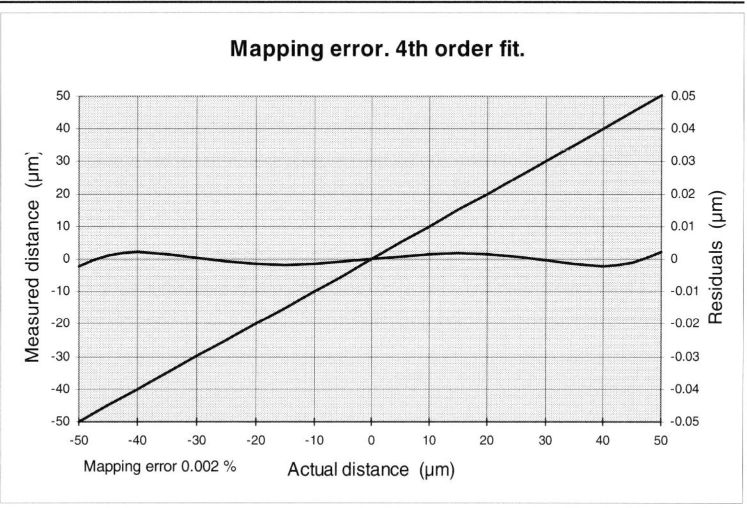

Fig. 2.7. 4th order fit

for an example see Fig. 2.6. Of course, if one were to use the higher order terms, an even better result could be achieved. This can be done easily in microprocessor based sensor systems or externally in the user's computer. It has been found that there is little to gain in going higher than fourth order, see Fig. 2.7.

Local Scale Factor Variation

Although a linearity error of 0.1 % of full scale is quite low, one can see that there are some steep local gradients in the residuals. These cause the local scale factor to differ significantly from the large amplitude scale factor which is usually quoted. That is, the scale factor varies according to the range over which it is measured, see Fig. 2.8. Typically if the linearity error is 0.1 % there could be a 1.5 % change in scale factor from one end of the range to the other, for example it could go from 0.995 to 1.01 over the range. The effect of this is that

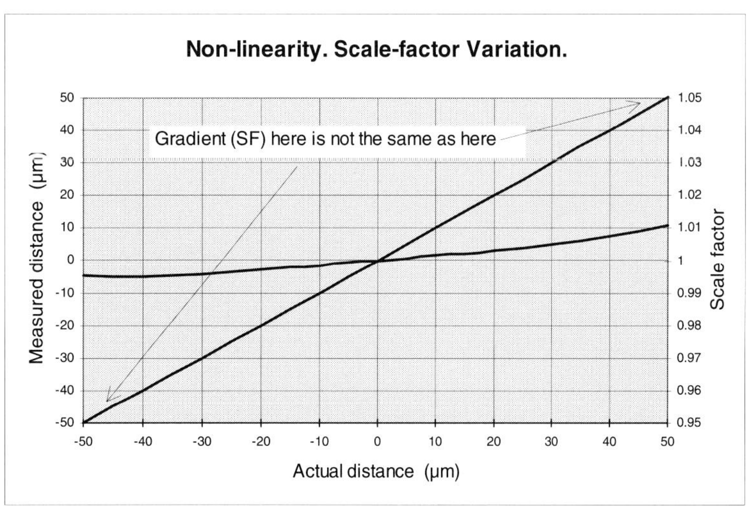

Fig. 2.8. Local scale factor variation

true motion corresponding to a measured 1 µm movement at one end of the range, say from 0 to 1 µm, would not be the same as the true motion corresponding to a measured 1 µm at the other end, say from 99 to 100 µm. The error would, of course, be within the quoted precision of the system but if the variation is known, it can be compensated.

Cosine and Abbe Errors

Measure the thing you want to measure

Ernst Karl Abbe (1840-1905) was a German mathematician and physicist who in 1866 was invited by Carl Zeiss to become his director of research. After the death of Zeiss in 1888, he set up the Carl Zeiss Foundation for research.

Abbe is famous for, among other things, his alignment principle which says: "when measuring the displacement of a specified point, it is not sufficient to have the axis of the probe parallel to the direction of motion, the axis should also be aligned with (pass through) the point"

These things come about by a misalignment of the motion and measurement axes. So far we have discussed the stage represented by Fig.2.2 with the sensor mounted on the X axis and motion considered along the X axis. A real world situation may be more like Fig. 2.9:

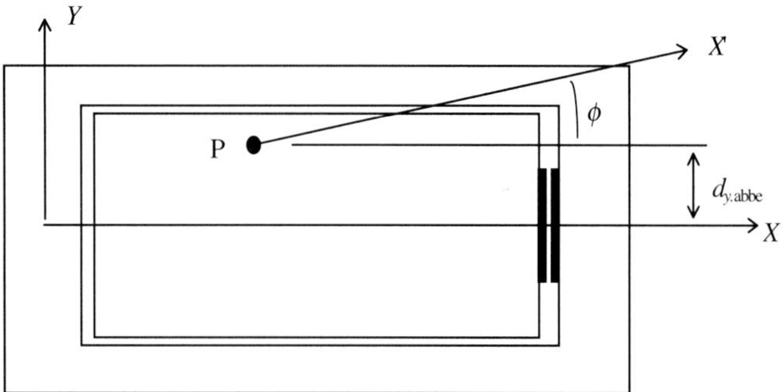

Fig. 2.9. Misaligned measurement

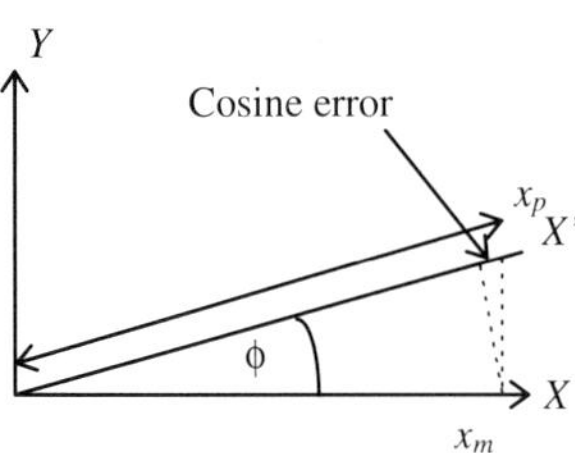

Fig. 2.10 Cosine error

Here the point P whose position is to be measured is offset from the X axis by a distance $d_{y.\text{abbe}}$. The true position x_p is also along an axis X' which is at an angle ϕ to the measurement axis X. This may happen if the stage mounting is misaligned. Now the measured position along the X axis x_m will be less than the true position, there is a *cosine error* given by

$$\delta x_{m.\cos\phi} = \frac{(1-\cos\phi)}{\cos\phi} x_m \qquad (2.7).$$

This is illustrated in Fig. 2.10. If ϕ is constant, then this manifests itself as an error in the scale factor that, in the case of a closed-loop stage, would be calibrated out. If the sensors are used alone however, care must be taken to ensure that ϕ is small.

As shown in Fig. 2.11, things get worse when ϕ varies as the stage moves (due to, say, yaw). Consider a motion along an arc from P to the true position x_p. Now there is an additional error given approximately by

$$\delta x_{m\phi.abbe} = d_{y.abbe}.\sin\delta\phi \qquad (2.8)$$

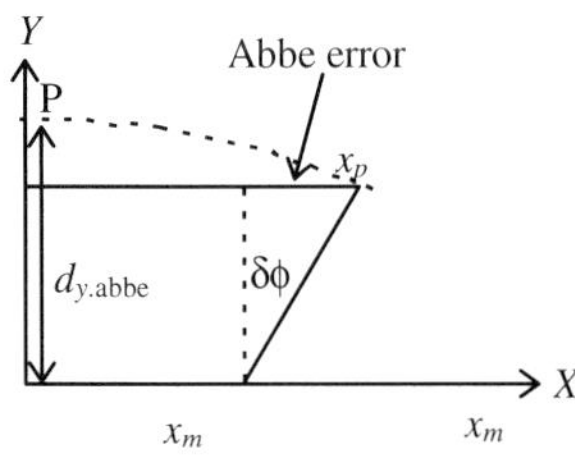

Fig. 2.11 Abbe error

where $\delta\phi$ is the total variation in ϕ as the stage moves. This is known as the *Abbe* error and $d_{y.abbe}$ as the *Abbe offset*. Obviously if $d_{y.abbe}$ is zero there is no Abbe error. The Abbe error cannot be calibrated out as it is not known where on the stage the user wishes the measured motion to be. It can however be compensated by the user if the variation in ϕ is known. This is discussed further in *Positioning*.

Both the cosine and Abbe errors are limitations to measurement trueness.

Measurement Resolution and Noise

We also worry a lot about these

Sensor noise manifests itself as a variation in the measured position about some mean value even when the stage being measured is not moving. Noise can be characterised by its frequency distribution (the noise spectrum), its amplitude distribution and the physical mechanism responsible for its generation.

Amplitude and Frequency Distribution

Noise is usually quoted as a rms (root mean square) value as this can be measured readily with standard equipment. Peak-to-peak noise levels are quoted by some vendors, but this information is not easily measured or interpreted, since with any noise distribution at some time there will be a large deviation from the mean. The noise amplitude distribution is important when looking at resolution, i.e. within what distance from the mean is a given percentage of the deviation? Usually Gaussian noise dominates and in this case the rms is equivalent to the

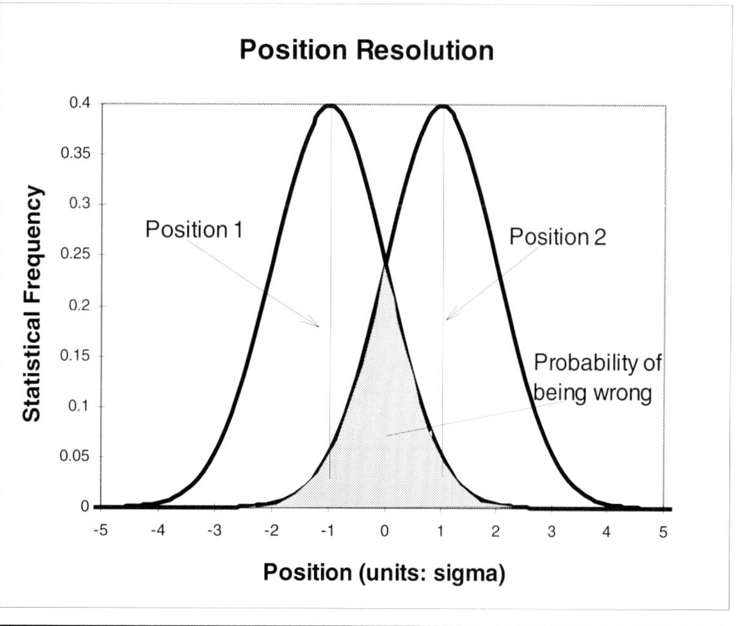

Fig.2.12 Resolving two positions

standard deviation, sigma: 68 % of samples taken will be within one sigma of the mean value.

The noise power spectrum as illustrated in Fig. 2.13 is a useful piece of information, since it can show up the underlying sources of noise - e.g. mains pick up, which is localised at 50 or 60 Hz, or high frequency components at say 10 MHz, which would be due to electro-magnetic interference (EMI).

In order to predict a measurement uncertainty within a system it is usually necessary to identify the different components of noise in the system and then add them together.

Sensor Noise Sources

There are many sources of sensor noise. Typically there will be internally and externally generated electrical noise and also externally generated mechanical noise. The following tables lists some of the more common noise sources and their properties.

Source	Properties	Comments
Internal Sources		
Amplifier noise	Gaussian. Bandwidth 0 to 5kHz	Usually the dominant source
Processor clocks	Distinct frequencies up to 50MHz.	Microprocessor based systems only. Amplitude minimised by design
Power-supply switching	Distinct frequencies around 50kHz to 100kHz	Minimised by design
External Sources		
Radio communications	Distinct frequencies 10s to 100s of MHz	Systems are well shielded against this. Care must be taken with sensor mounting when transmitters are close.
Mains-borne spikes from motors etc.	Essentially random frequencies up to several MHz	Power line filtering is used to minimise this problem.
Acoustic noise from air-handling systems, rock concerts etc.	10s to 1000s of Hz	Must be minimised by careful sensor mounting.
Ground vibrations due to traffic etc.	10s to 100s of Hz	Must be minimised by careful sensor mounting.

Table 2.3 Noise Sources

Combining noise source

The overall effect of noise sources will depend on the nature of the noise and the measurement bandwidth (the subject of bandwidth is discussed in more detail in Chapter 3, Servo Control). The spectrum example of Fig. 2.13 shows a system with 500 Hz bandwidth, but for illustration the spectrum from 1 Hz to 100 MHz is shown. Most of the noise in the band is 'white' Gaussian noise characterised by a nominally flat spectrum from 1 to 500 Hz. The example level is 15 pm·Hz$^{-1/2}$. The contribution to the overall measurement noise from this Gaussian component, $\delta x_{m.ng}$, depends on the square root of the bandwidth

$$\delta x_{m.ng} = \delta x_{m.ndens} \cdot \sqrt{B_m} \tag{2.9}$$

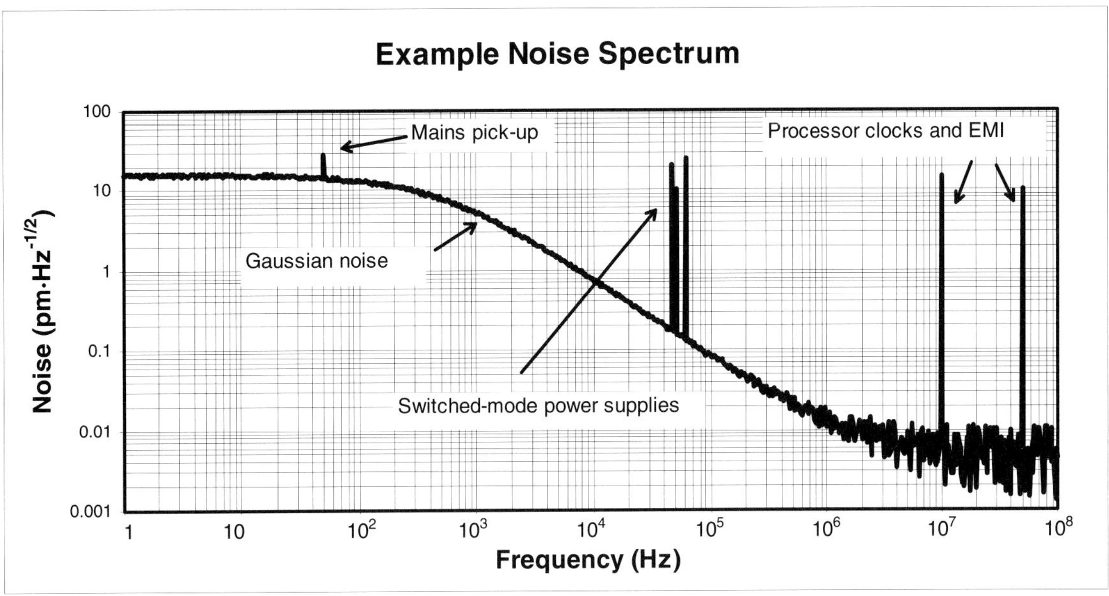

Fig.2.13. Noise spectrum

where $\delta x_{m.\text{ndens}}$ is the Gaussian noise density (in this case 15pm.Hz$^{-1/2}$) and B_m is the bandwidth (500 Hz).

The spectrum example shows peaks at around 50 kHz which could be due to interference from switched mode powers supplies and further peaks at 10 MHz and 50 MHz which could be due to processor clocks or external interference. Most stages cannot physically move this fast so these components can be ignored when assessing the noise equivalent displacement of the stage. In practice of course one would never look at the sensor output with such a wide bandwidth.

There is also a peak shown at 50 Hz. This could be due to pickup from the mains supply on the sensor circuitry or it could be a real stage motion. It cannot be ignored and must be added in. An rms addition is used

$$\delta x_{m.\text{nsens}} = \sqrt{\delta x_{m.\text{ng}}^{\,2} + \delta x_{m.\text{n}50}^{\,2}} \qquad (2.10)$$

where $\delta x_{m.\text{n}50}$ is the rms value of the 50 Hz component. It should be noted that the value of $\delta x_{m.\text{ng}}$ depends on the bandwidth selected but $\delta x_{m.\text{n}50}$ will be the same for all bandwidths above about 50 Hz.

Resolution

The position resolution of a purely analogue sensor or system is usually defined by the noise level. However if the position information is transferred digitally, which is usually the case, then the number of bits used to transfer the information must also be considered. In high precision systems, Queensgate uses single precision floating point number format to transfer information. This has a range of seven decimal digits giving a resolution of 0.1×10^{-6} or 10 pm in a system with 100 µm range, which is far less than the analogue noise for systems with bandwidths above a hertz or so. Less precise systems use 16 bit integers to code information giving a digital resolution of 1 part in 65 536 or 1.5 nm in a 100 µm range system. This can be significant, especially when the signal being digitised has intrinsically low noise.

If N is the number of bits used to transfer information, then we have an effective extra noise source (quantisation noise). The rms value of this noise source is

$$\delta x_{m.\mathrm{nquant}} = \frac{0.29 d_{xm.\mathrm{max}}}{2^N} \tag{2.11}$$

where $d_{xm.\mathrm{max}}$ is the full measurement range. The derivation of this is given in the Appendix to this chapter. The total noise can be derived from the sensor noise given by equation 2.10 and the quantisation noise given by 2.11

$$\delta x_{m.n} = \sqrt{\delta x_{m.\mathrm{nsens}}^2 + \delta x_{m.\mathrm{nquant}}^2} \tag{2.12}.$$

Measurement Repeatability and Reproducibility

Measurement repeatability is defined as the closeness of the measured displacements when the same true half range displacement is applied to the sensor repeatedly, under the same operating conditions and in the same direction. Repeatability does not include hysteresis or drift.

Measurement reproducibility is defined as the closeness of the measured displacements when the same true half range displacement is applied to the sensor repeatedly, approaching from both directions. Reproducibility includes hysteresis, dead band, drift and repeatability.

These definitions are derived from ISO 5725.

Sensor Hysteresis

It should be noted that though capacitance sensors are intrinsically hysteresis free, this may not be true of other sensors. For example resistive strain-gauges can exhibit high intrinsic internal hysteresis as well as drift.

Intrinsically a capacitance sensor has zero repeatability, zero hysteresis and zero dead band error: there is a direct relationship between the capacitor gap and the electrical capacitance. The only contribution to reproducibility error will be drift. In the real world however, the sensor can rarely be mounted at the same place as the 'true' displacement to be measured is applied, see Fig.2.9. This can lead to hysteresis errors, but as they are concerned mainly with positioning rather than measurement they are dealt with under *Position Repeatability* and *Position Reproducibility*.

If the plates of a capacitance sensor are held at a fixed distance apart, the measured position may vary with time: a_{x0} in equation 2.4 may change. This drift is usually associated with environmental changes, such as temperature or humidity which can affect the signal conditioning electronics or the capacitors directly. The errors contribute to the precision of the measurement, δx_{mR}. Careful design of the capacitors and the electronics minimises the environmental effects, but as they can be predicted by monitoring the environment, temperature and humidity coefficients are quoted as separate contributions to precision so their effects can be minimised if required. See Chapter 5, *Capacitance sensors, Environmental effects* for further discussion of this.

POSITIONING

Where you are, where you think you are and where you want to be

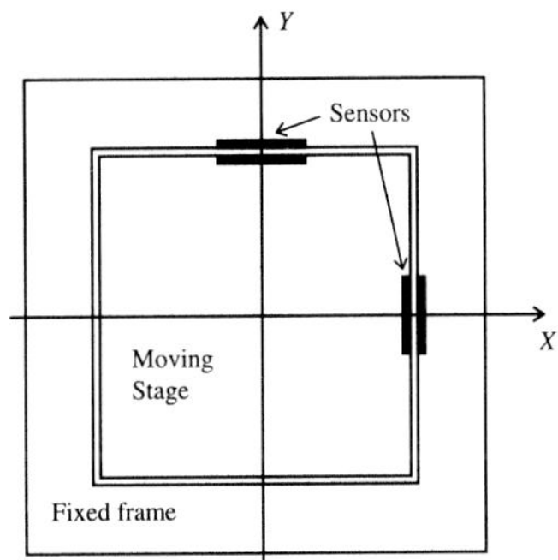

Fig. 2.14. A simple XY stage

When considering positioning accuracy it is useful to talk around a closed loop positioning stage where motion is produced with piezo translators and sensed with capacitance micrometers. To make life more complicated, we will consider a two dimensional XY stage shown diagramatically in Fig. 2.14. Here the stage is moved in the *X* and *Y* direction with piezo translators (not shown, but acting along the *X* and *Y* axes) and movement measured with capacitance sensors mounted on the *X* and *Y* axes as shown.

Usually the desired motion will be determined by the user in conjunction with a computer that sends *XY* position commands to the controller. The motion measured by the sensors in the two axes is monitored by the computer and also fed back to the controller which moves the stage to minimise the difference between the sensed motion and the command. Fig. 2.15 shows a simple control loop and Fig. 2.16 various positions.

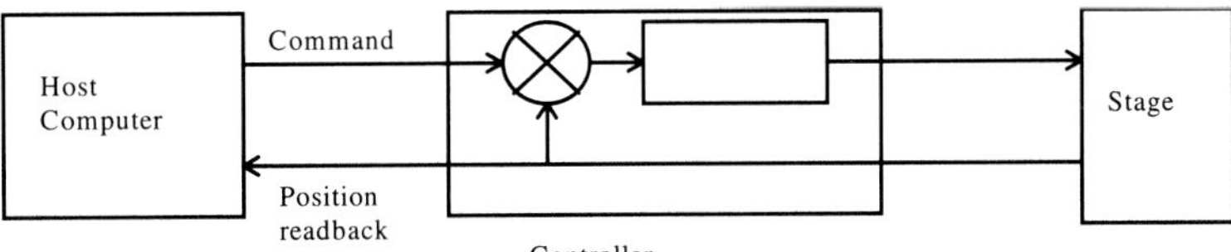

Fig. 2.15. Simple control loop

Fig. 2.16 is an extension of Fig. 2.3 with some additions.

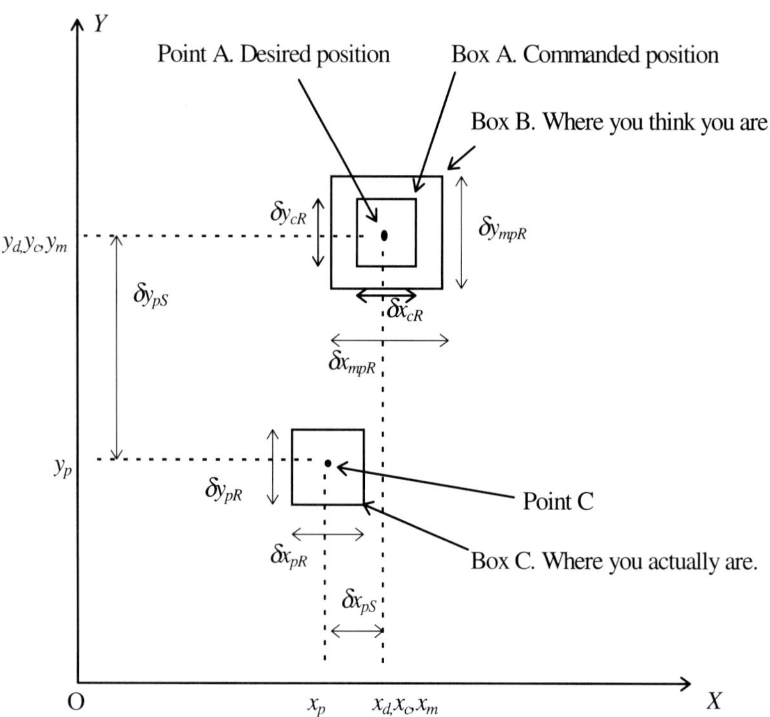

Fig.2.16 Positions

Consider a movement starting from an origin 'O'. Point A in Fig. 2.16 represents a desired position and Box A the position actually commanded. Ideally Box A would be a point the same as Point A, but as will be shown later this may not be the case as positions can only be coded with a certain precision, δx_{cR}, δy_{cR}

The larger box around position A, Box B, represents where you think the stage has moved to, using the stages built-in sensors. The size of the box represents the sensor *measured position precision* which has components δx_{mpR} and δx_{mpR}. This is not quite the same as the measurement precision defined in Fig.2.3, as now there may be real random motions of the stage which may make the box bigger. In a closed-loop system the box is centred on the *commanded position*: if it were not the servo would see the displacement and correct it.

If now the stage is observed using an external perfect noiseless position measuring device, the true position observed would be somewhat different again. This is represented by the third box, C. Again this is a box because the external device will see any random motions produced by the positioning system. Also if the same position is commanded several times, the resultant position may not be the same in every case. The distance between the desired position and the true position represents the *position trueness* which has components δx_{pS} and δy_{pS}. The size of Box C represents the *position precision* with components $\delta x_{pR}, \delta y_{pR}$. The position trueness and position precision make up the *position accuracy*:

$$\delta x_{pA} = \delta x_{pS} + \delta x_{pR} \qquad (2.13)$$

$$\delta y_{pA} = \delta y_{pS} + \delta y_{pR} \qquad (2.14).$$

When describing the system performance we must consider the ability of the system to move to a required position and also the ability of the system to tell the user where it has moved to. As can be seen, these may not be the same. In the same way as measurement trueness and precision are made up of other parameters, so it is with position trueness and precision:

Property	Constituent parameters	Dealt with in sub-section:
Position trueness	Mapping trueness	Position Linearity and Mapping
	Abbe error	Cosine and Abbe Errors
	Cosine error	Cosine and Abbe Errors
Position precision	Mapping error	Position Linearity and Mapping
	Resolution	Position Resolution and Noise
	Noise	Position Resolution and Noise
	Reproducibility	Position Reproducibility
	Repeatability	Position Repeatability

Table 2.4 Position trueness and precision

All the parameters that affect measurement also affect positioning. That is not surprising as in a closed loop system it is the performance of the measuring system that dominates the positioning accuracy: if the measuring system can say where the stage is with certainty then the servo controller can make this position equal to the required position (within limits - see Chapter 3 Servo Control).

The constituent parameters can be broken down further in exactly the same way as with measurement, Fig.2.4 still applies.

Position Linearity and Mapping

In a simple control loop, the measured position is compared with the command position and the stage moved to null the difference, that is the measured position will be made equal to the command. If, then, the measured position is a linear function of the true position, the true position will be a linear function of the command. This is discussed in detail in Chapter 3, Servo Control. As has been discussed in *Measurement linearity and mapping*, the measured position may not, however, be a linear function of the true position but may best be described by a higher order power series. In this case the true position will not be a linear function of the command but will be described by another power series, the *positioning mapping function.*

The input to the positioning mapping function is the desired position, Point A, and the output is the true position, Box C. Although they are related, it should be noted that in a closed loop positioning stage the positioning mapping parameters are not identical to the measurement mapping parameters. Equation 2.4, copied here, relates the measured displacement x_m to the true displacement x_p

$$x_m = a_0 + a_1 x_p + a_2 x_p{}^2 + a_3 x_p{}^3 + a_4 x_p{}^4 \ldots\ldots\ldots$$

(2.15).

In the closed loop system x_m is made equal to the input command, x_c, so the equation must be reversed to get the true position x_p. Now we have

$$x_p = b_0 + b_1 x_c + b_2 x_c{}^2 + b_3 x_c{}^3 + b_4 x_c{}^4 \ldots\ldots\ldots$$

(2.16).

Similar equations exist for y.

For the linear approximation we can define the positioning scale factor, b_1:

$$b_1 = \frac{1}{a_1}$$

(2.17).

This is simple enough for a linear fit to the actual transfer function but reversing a fourth order mapping function can be more complex! This is done in Chapter 3, Servo Control, but in general the user does not have to worry about it: for stand alone sensors the position to measurement mapping function is supplied and for positioning stages the command to position function is supplied: it all gets calibrated out.

Position Resolution and Noise

Position noise manifests itself as a variation in the true stage position about some mean value even when the commanded motion is zero. It is a contribution to the size of Box C.

Contribution from Sensor.

Just as the measurement trueness determines the positioning trueness, so sensor noise and resolution affects positioning noise and resolution. With a stationary command, the sensor noise will generate an error signal that will be interpreted by the control loop as a command. It will move the stage to null the error signal thus making the noise signal an actual displacement. In a purely analogue system there will be a contribution to position noise of $\delta x_{p.nsens}$ given by equation 2.10. In a digital system the contribution will be given by equation 2.12.

Command Resolution.

Starting at the beginning, the system must be able to command the desired motion: ideally box A should be zero size. Usually the requirement for a movement is determined by the user and a computer and the position information will be transferred to the controller as a binary number. The number of bits used to represent the position determines how well this can be done. As with sensors, in high precision systems, Queensgate uses single precision floating point number format to transfer information. This has a range of seven decimal digits giving a resolution of 0.1×10^{-6} or 10 pm in a system with 100 µm range. This is unlikely to be a limitation but if fewer bits are used an error could be introduced. For example if the full range of a stage and controller system is 100 µm and the command information is coded as a 16 bit number, the smallest step that can be recognised is

$$step = \frac{100 \mu m}{2^{16}} = 1.5 nm \tag{2.19}.$$

If a motion of 3 nm is required then the system can code this exactly and there is no error at this point. If 4 nm is required the nearest command is 4.5 nm and there will be a 0.5 nm error. In general there is an uncertainty of plus or minus half the least significant bit value or a noise equivalent displacement as given by equation 2.11. For the command, this is designated $\delta x_{c.nquant}$.

Piezo Drive Noise.

Once the position of the stage has been measured and compared with the command position, the resultant difference signal is used to generate a drive voltage to be applied to the piezo actuators. Noise will be introduced by this process so even if the measurement system were perfectly noiseless, the stage would still have a noise motion due to this.

The effect of noise introduced at this point in the loop is somewhat different to that introduced by the measurement micrometer in that it is at least partially servoed out. The ability of the system to servo out the drive noise depends on the bandwidth set: the higher the bandwidth the better the contribution is servoed out. The piezo drive noise actually *decreases* with increasing bandwidth! Compared with the sensor noise however, the piezo drive noise contribution is small, so its reducing contribution at high bandwidth is swamped by the increase in sensor noise.

This drive contribution is designated $\delta x_{p.ndrive}$ and in practice represents the 'noise floor' of the system: even at zero system bandwidth there will be stage motion at this level.

Mechanical and Acoustic Noise

External mechanical inputs such as ground vibration and acoustic noise will cause the stage to move. The effects of these inputs can be minimised by designing the stage so that it is stiff, which will maximise immunity to vibration, and by careful stage mounting.

Vibration that does make its way into the stage will be servoed out if the vibration is within the system bandwidth. Mechanical noise can be quite high in some environments and in this case having a high system bandwidth *reduces* the effect of the noise input.

Position Reproducibility

Position *reproducibility* is defined as the closeness of the generated displacements when the same true half range displacement is commanded repeatedly, approaching from both directions. Reproducibility includes hysteresis, dead band, drift and repeatability.

Hysteresis

Hysteresis manifests itself as a difference in true position when the required position is approached from different directions. The simplest form of hysteresis is known as backlash. This is usually found in devices operated by a lead screw - when the direction of motion is changed there is a small *dead band* when the screw turns, but no motion is produced - see Fig.2.17. Piezo-electric devices (and flex hinges) do not have a dead band but exhibit a more complex form of hysteresis where the position depends to some extent on previous positions. This is discussed in Chapter 6 Piezos, but Fig. 2.18 shows a typical 'hysteresis loop'.

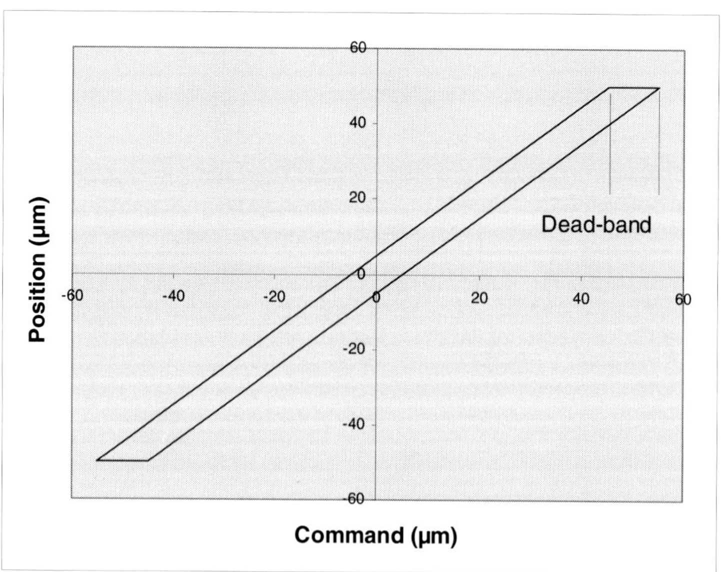

Fig. 2.17 Dead-band

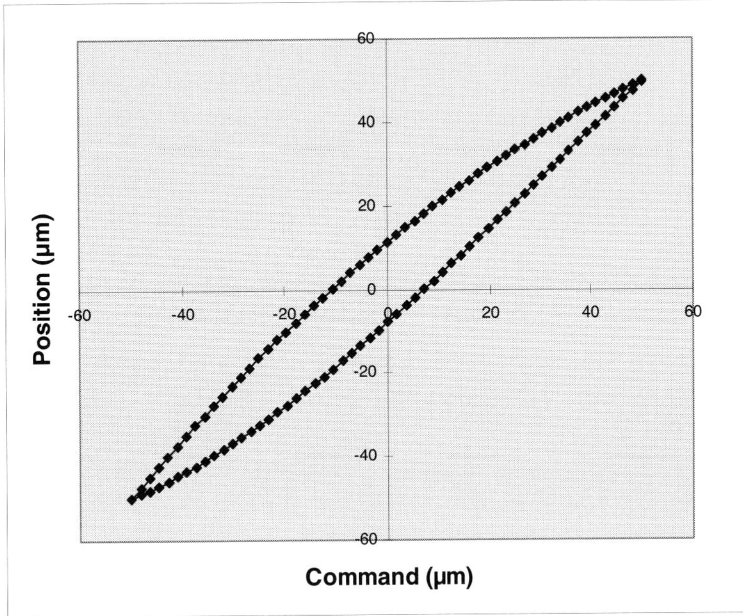

Fig.2.18 Hysteresis Loop

In a closed-loop system where the position of the stage is monitored with a capacitance micrometer, hysteresis is taken out and the stage performance depends only on the sensor. That is in an ideal world. In practice the sensor can never be in exactly the same place on the stage as the position to be controlled: the Abbe offset can have an effect (see page 12). In all Queensgate stages the sensors are mounted on the motion axes. Motions along that axis are thus well controlled, but if the point of interest is offset from the axis, then rotation in the ϕ direction can affect the position. The ϕ rotation is uncontrolled and is governed

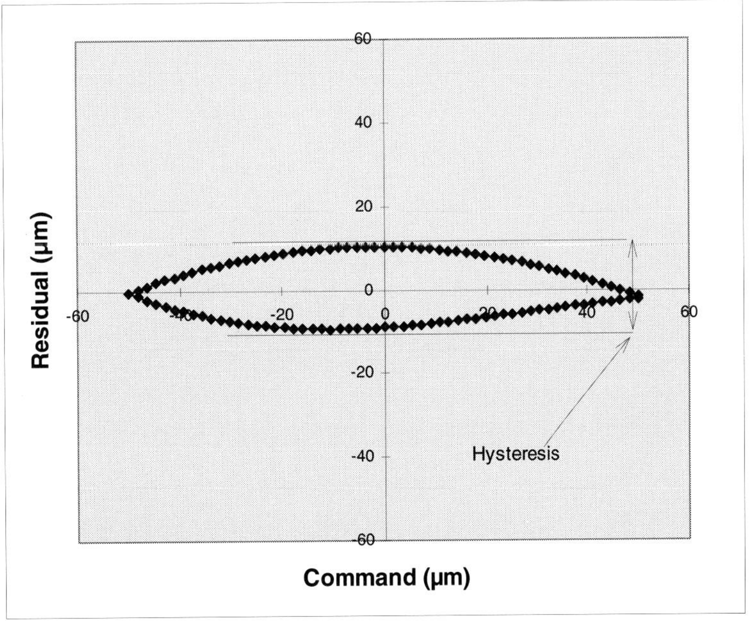

Fig.2.19 Hysteresis Measurement

by asymmetries in the moving stage mounting hinges and off-axis forces from the piezo actuator so the Abbe error could have a hysteresis component. This is minimised by design but is measured and quoted by Queensgate.

The most telling measure of hysteresis is to scan the stage backwards and forwards several times, fit a linearisation polynomial to the data (generally this will be a straight line) and plot the residuals. Any hysteretic tendency is then clearly visible and is quoted as a fraction of the full scan range. Fig.2.19 illustrates the procedure for an open-loop piezo showing large hysteresis. Some manufacturers quote a random bi-directional repeatability which measures the variation as an rms deviation when the system is offset by varying amounts of random direction and amplitude. This can however produce a figure which flatters, masking a large hysteresis loop.

Position Repeatability

Repeatability is defined as the closeness of the generated displacements when the same true half range displacement is commanded repeatedly, under the same operating conditions and in the same direction. Repeatability *does not* include hysteresis or drift. This is often used by manufacturers of lead screw type devices, to separate the accuracy of the motion from the backlash. It is often called *uni-directional repeatability*.

Summary of contributions to position accuracy

Some of the terms that make up positioning precision and trueness are illustrated graphically in Fig. 2.20.

This figure shows various positions for a closed-loop system plotted against the desired position, which is assumed to be scanning over the full range in both directions. In an ideal system all the lines would be on top of one another but various errors spread them out. The errors have been grossly magnified for clarity (especially hysteresis and non-linearity which would be very close to zero in a closed-loop stage) and they all should show quantisation steps, apart from the desired position, but the drawing is too complicated as it is!

As can be seen from this plot, the position precision bar may not be centred on the true position as non-linearity and hysteresis may push the true position off centre.

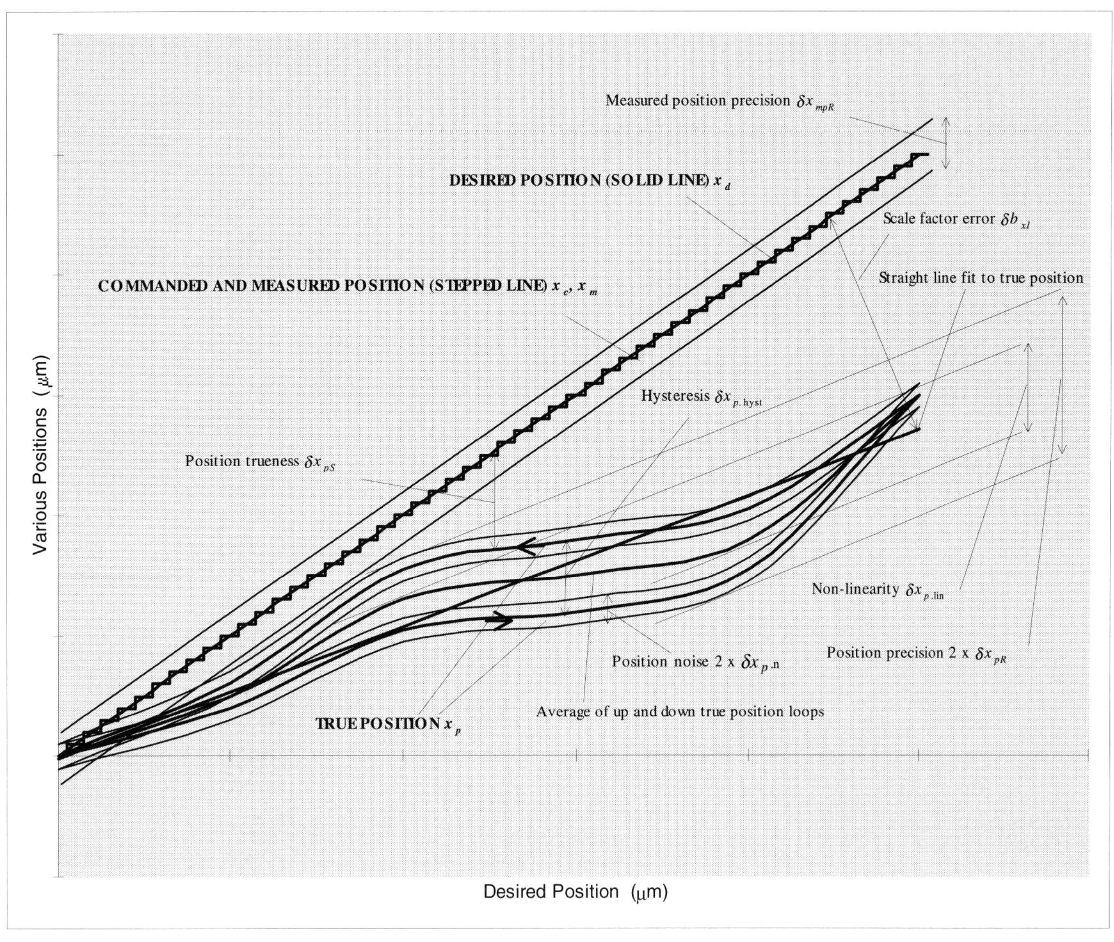

Fig. 2.20 contributions to precision and accuracy

APPENDIX.

Derivation of Quantisation Noise

The finite number of bits used to encode a signal gives an uncertainty in the value of that signal. This can be treated as an extra noise source.

Consider a signal that does not vary in an arbitrary time interval T. If the N bit representation of the true value of the signal is n, the true value of that signal may fall between two values $n\text{-}LSB/2$ and $n\text{+}LSB/2$ where LSB represents the value of the least significant bit of the number. We cannot know where in between these two values the true value lies: any value is equally probable.

Given that any value between the limits is equally possible, the value can be described by a straight line from $-LSB/2$ to $+LSB/2$ as illustrated in Fig. 2.21, this gives all values equal weighting in the simplest way.

This line can be represented by

$$y = \frac{t}{T} - \frac{1}{2} \tag{2.20}$$

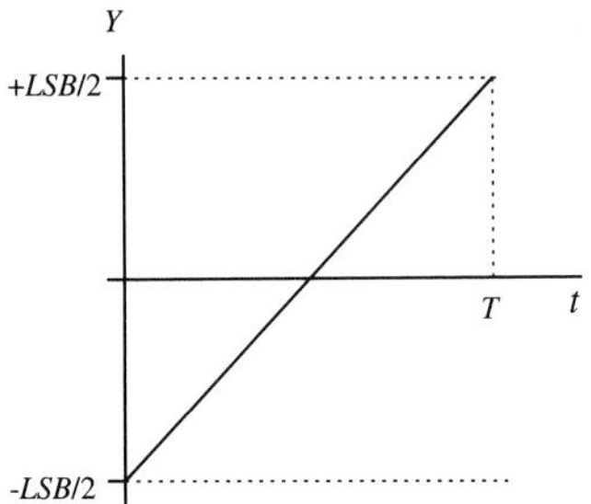

Fig. 2.21. Possible values

where the value of LSB is assumed to be unity. We can now find the rms value of this function from 0 to T to define an equivalent noise

$$rms = \sqrt{\frac{1}{T} \int_0^T \left(\frac{t}{T} - \frac{1}{2} \right)^2 dt} \tag{2.21}$$

$$rms = \sqrt{\frac{1}{T} \left[\frac{t^3}{3T^2} - \frac{t^2}{2T} + \frac{t}{4} \right]_0^T} \tag{2.22}$$

$$rms = \sqrt{\left(\frac{1}{3} - \frac{1}{4} \right)} = 0.29 \tag{2.23}.$$

This is the rms noise equivalent to the least significant bit, so we can define an actual quantisation noise as

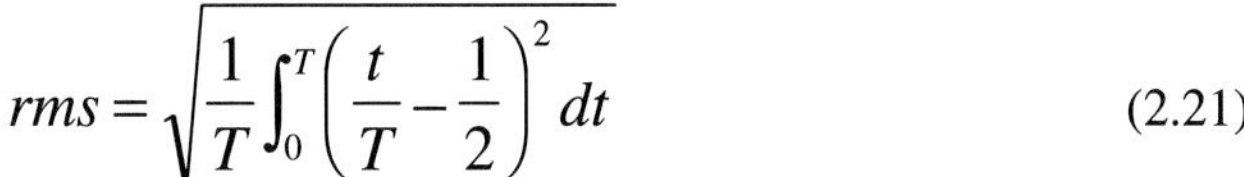

$$\delta x_{m.\text{nquant}} = \frac{0.29 d_{xm.\text{max}}}{2^N} \tag{2.24}$$

where $d_{xm.\text{max}}$ is the maximum measurement range.

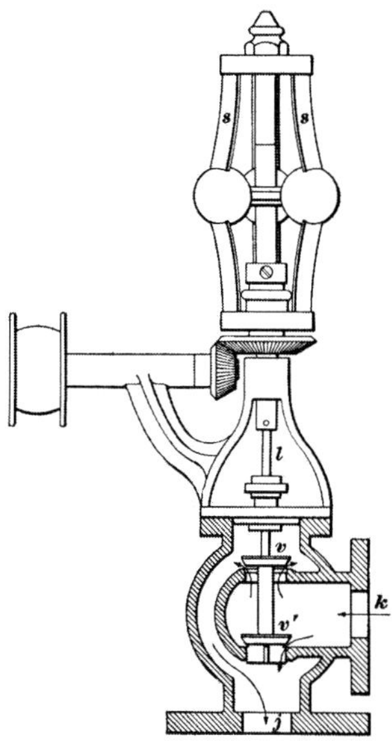

A steam governor. An example of mechanical feedback control

Many text books have been written on control theory. It is not the purpose of this section to write another one, rather to describe in qualitative terms the systems used by Queensgate and show how various control loop parameters affect performance.

In this chapter:

STATIC PERFORMANCE

Open-loop control

It is the aim of any position control system to make the output position reflect an input command. This is illustrated by the simple 'open-loop' controller of Fig.3.1.

Fig. 3.1 Open loop control

Classical servo control theory is dominated by the use of transfer functions in terms of the complex frequency 's' and their manipulation using Laplace transforms. It is not necessary to have a knowledge of these techniques for an understanding of this chapter, one can replace s with jω to get an idea of the shape in the frequency domain of the functions used. Here ω is the angular frequency, 2πf, and j the square root of minus one,

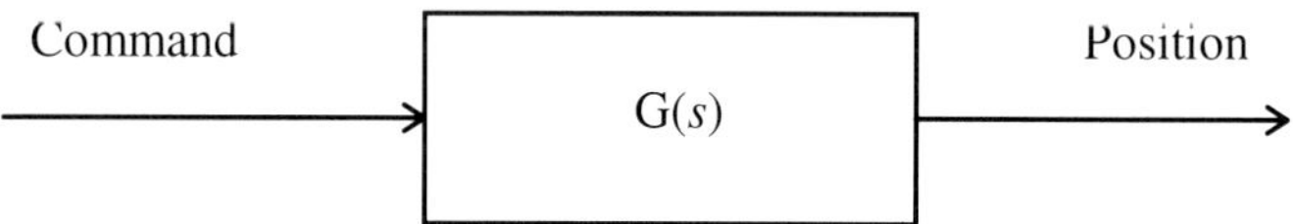

Here position is related to the command by the 'transfer function' $G(s)$ where s is the complex frequency.

The command will be in the form of a voltage in an analogue system or a binary number for a digital system. For example in an analogue system an input of zero volts will set the output to the nominal zero μm position, an input command of 10 V may move the stage 50 μm to the 50 μm position, an input command of -5 V may move it to the -25 μm

It should be noted that Queensgate's NPS range of digital controllers has a floating point digital command option: then the input command really is in µm.

position and so on. Such a system would have a scale factor of 5 µm·V⁻¹. In a corresponding 16 bit digital system +32 767 would give +50 µm, -32 768 would give -50 µm. Obviously the actual scale factor will depend on the range of the system being considered. Queensgate makes stages with full ranges of 10 µm to more than 100 µm, so to keep things general we will consider the input command to be in µm. What this corresponds to in volts or numbers depends on the actual stage and controller.

With the command and position both in µm, the transfer function G in Fig. 3.1 should be unity. It is shown as a function of 's' as it will depend on the frequency of the input command: more of this later. In practice G will be a non-linear function of the input command and will be hysteretic if piezos are providing the motion. It may also vary with temperature and time and have an offset added to it that may also vary with temperature and time. All in all the position will be only loosely related to the input command! Let us expand Fig. 3.1 to include some of the errors:

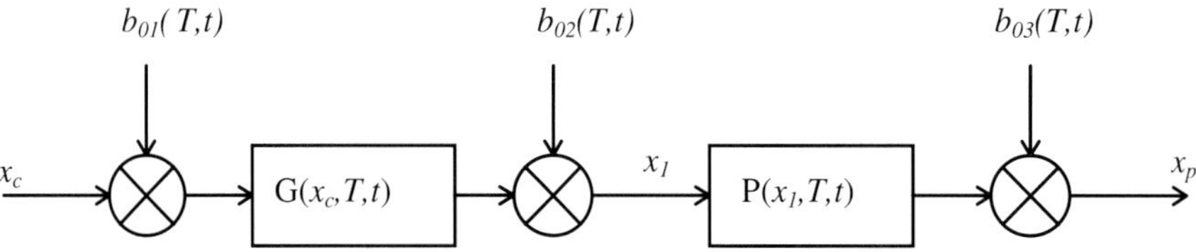

Fig. 3.2. Open-loop with offsets

b_{01}, b_{02}, b_{03} are offsets added in at various points in the controller, x_c is the command and x_p the true position (see equation 2.16). The gain term G is now a function of the input command illustrating non-linearity and also a function of temperature T and time t. The frequency dependence, s, has been ignored for now. P represents the piezo actuator which is non-linear (and hysteretic) in the intermediate variable x_1 and is also a function of temperature and time. In this example the time dependence is included to show the effects of drift, it does not relate to time variation of x_c: the command is considered to be static for the moment. The offsets b_{01}, b_{02} are electronic but b_{03} could be mechanical, caused for example by thermal expansion of the piezo actuator or the metalwork between it and the output measurement point. We thus have

$$x_p = b_{03} + P\left[b_{02} + G\left(x_c + b_{01}\right)\right]$$

(3.1).

The x, T and t dependencies of the various parameters have been omitted from this equation for clarity. Given that G and P are known, one can assess the effect of the various offsets. The input offset b_{01} is indistinguishable from the command x_c and is multiplied by G and P in the same way as the command; the intermediate offset b_{02} is only multiplied by P and the output offset b_{03} appears directly on the position. Without putting numbers in, it is obvious that all these temperature dependencies and non-linearities are undesirable.

Closed-loop Control

A Simple System

A lot of these nasty features can be eliminated with closed-loop control. The principle of closed-loop control is to measure the true output and compare it with the required output. If the two are not the same, then change the output until they are! The classic way of doing this is shown in Fig. 3.3.

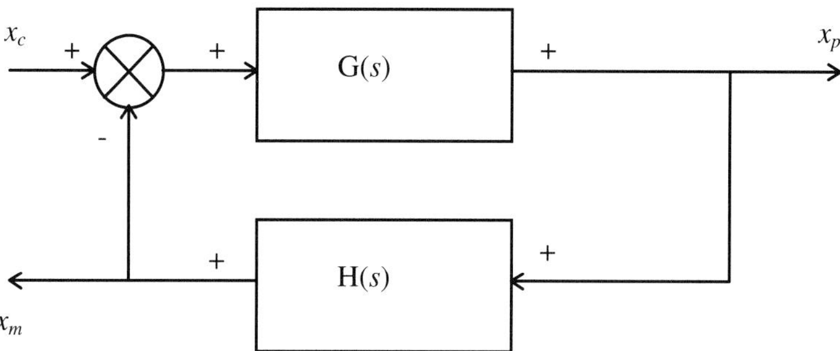

Fig. 3.3 A Simple Closed Loop

H(s) represents the transfer function of the sensor system. By examination of Fig. 3.3

$$x_p = G\left(x_c - H.x_p\right) \tag{3.2}$$

from which

$$x_p = \frac{G.x_c}{1 + G.H} \tag{3.3}.$$

Again the s dependence has been ignored for now. If G is made very large, then equation 3.3 reduces to

$$x_p = \frac{x_c}{H} \tag{3.4}.$$

This is now completely independent of G so all its temperature, time (drift) and non-linear effects go away. Of course H will be unity as the measurement should equal the true position, in which case

$$x_p = x_c \qquad (3.5)$$

which is exactly what is required. Imperfections in H will come through as errors but if the sensor is a capacitance micrometer, H will be non-hysteretic, linear and relatively drift free: see Chapter 5, Capacitance Sensors.

A System with Errors

The beneficial effects of closed-loop control can be seen explicitly by closing the loop on Fig. 3.2, the system with errors.

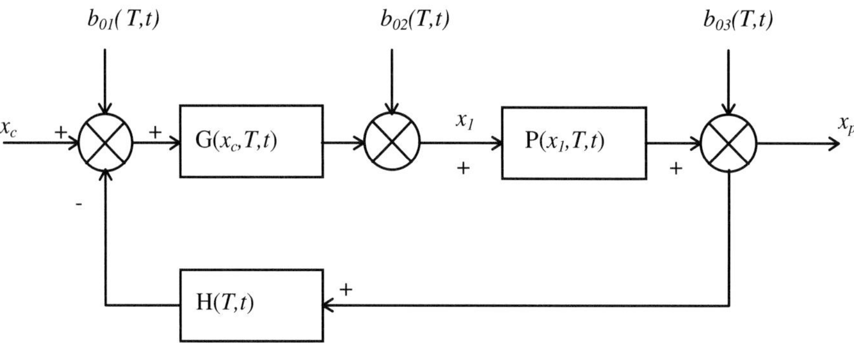

Fig. 3.4 Closed loop with Errors

H is shown as a function of temperature and time but it must be remembered that this is a very weak function for capacitance sensors. Now we have

$$x_p = b_{03} + P\left[b_{02} + G\left(x_c + b_{01} - Hx_p\right)\right] \qquad (3.6)$$

giving

$$x_p = \frac{b_{03}}{1 + HPG} + \frac{Pb_{02}}{1 + HPG} + \frac{PG\left(x_c + b_{01}\right)}{1 + HPG} \qquad (3.7).$$

As will be shown, G can be made effectively infinite which means that the effect of b_{03}, b_{02} and P disappear completely. If H is unity, we are left with

$$x_p = x_c + b_{01} \qquad (3.8).$$

It is thus impossible to remove the effect of the input offset, which is not surprising as it is indistinguishable from the command.

In summary, these exercises show that the static performance of the servo control loop is determined by the measurement system and not the piezos etc. In an open-loop system all the drift, non-linearity and hysteresis of the piezos are present on the output. In a closed-loop system they are not.

The Integrator

The closed-loop system relies for good offset elimination and high linearity on having a very high value for G. This can be achieved by making G an integral term, that is the output is the time integral of the input. This is illustrated in Fig. 3.5. The input to the integrator is shown as a 'top hat' rising from zero to 1 V at time 4 s and returning to zero at 14 s. The integral of this is a ramp rising to a maximum value (in this case 10 V) and staying there when the input returns to zero.

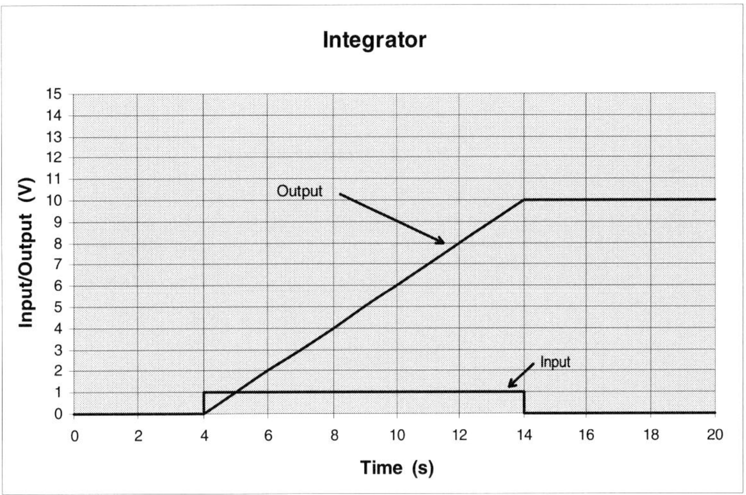

Fig. 3.5. Integrator action

If the input were to remain at its non zero value, ignoring obvious limitations like voltage headroom or computer number overflow, the output would ramp to infinity after an infinite time. The gain is thus infinite at dc (note that a steady input can only be said to be dc after an infinite time). Alternatively one can see that the output can be finite even with zero input, i.e. infinite gain.

The transfer function of an integrator is

$$G_{\text{int}} = \frac{1}{\tau s} \tag{3.9}$$

where τ is a time constant governing the time it takes for the output to get to a certain value. For a step input it is the time taken for the output to equal the input (1 s in the example). Replacing s with $j\omega$ shows that the gain goes to infinity at zero frequency (dc).

By using an integrator most *steady state* errors and non-linearities are thus eliminated. In practice an analogue integrator will not have infinite dc gain as it will be built around an operational amplifier having finite, though large, dc gain so a rising ramp would stop at this value times the input (again ignoring headroom limitations). Also an analogue integrator cannot hold an output indefinitely when the input is zero, due to leakage. A digital system can have infinite dc gain as a digital algorithm can ramp up for ever (ignoring numerical overflow) and can hold an output value indefinitely.

Static Mapping

The static performance of a closed-loop system depends on the performance of the position sensor. The mapping function for true position to measured position was defined by equation 2.4

$$x_m = a_0 + a_1 x_p + a_2 x_p^2 + a_3 x_p^3 + a_4 x_p^4 \ldots\ldots$$

$$(3.10).$$

The measured position will be made equal to the commanded position by the servo so equation 3.10 can be written

$$x_c = a_0 + a_1 x_p + a_2 x_p^2 + a_3 x_p^3 + a_4 x_p^4 \ldots\ldots$$

$$(3.11).$$

This needs to be reversed to get x_p as a function of x_c, giving

$$x_p = b_0 + b_1 x_c + b_2 x_c^2 + b_3 x_c^3 + b_4 x_c^4 \ldots\ldots$$

$$(3.12)$$

which was introduced as equation 2.16.

If one set of coefficients are known, the reverse set can be found using the formulae in the appendix to this chapter. It should be noted that Queensgate closed-loop position stages are calibrated directly in terms of true position as a function of command, i.e. the 'b' coefficients are measured directly, though generally only b_1 is of interest due to the high linearity of the capacitance sensors. Equation 3.12 then gives the mapping from command to position. Alternatively the 'a' coefficients can be derived and the command required to get a certain position determined using equation 3.11.

Internal Mapping Correction

For convenience we can write the mapping functions equations 3.11 and 3.12 as matrix operator equations

$$x_c = AX_p \qquad\qquad (3.13)$$

$$x_p = BX_c \qquad\qquad (3.14)$$

where $\mathbf{A}$ is a row vector of 'a' coefficients, $\mathbf{X_p}$ is a column of powers of x_p and similarly for $\mathbf{B}$ and $\mathbf{X_c}$. $\mathbf{A}$ and $\mathbf{B}$ are the reverse of each other, which can be written

$$\overline{A} = B \tag{3.15}$$

$$\overline{B} = A \tag{3.16}.$$

Given that $\mathbf{B}$ is determined by calibration, it can be applied to the sensor reading *before* comparison with the command to linearise the motion with respect to command. This is done in Queensgate's range of digital controllers: the principle is illustrated in Fig. 3.6.

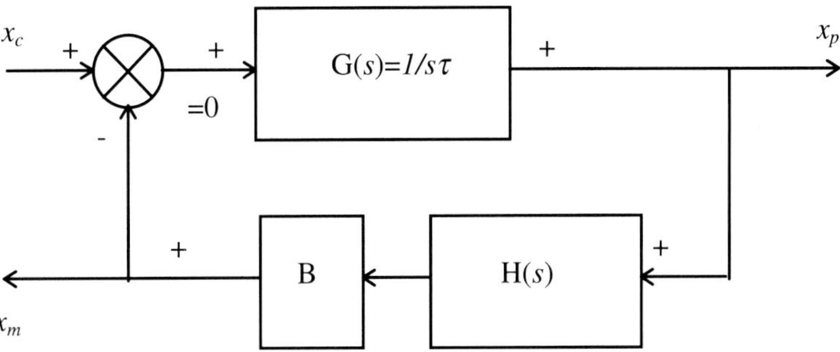

Fig. 3.6 Mapping Correction

It may at first sight seem that it should be the $\mathbf{A}$ mapping operator that is applied: it is not. The argument goes something like this:

1. Consider the stage moving linearly with respect to the command;

2. the sensor output can be predicted by applying $\mathbf{A}$ to the stage motion, the output may be non-linear;

3. if $\mathbf{A}$ transforms a linear motion to something non-linear, then $\overline{A}$ will transform the non-linear sensor output back to something linear;

4. the linearised sensor output is made equal to the command so the command is now linear with respect to the motion;

5. if the command is linear with respect to the motion, then the motion is linear with respect to the command as originally postulated;

6. $\overline{A} = B$ so $\mathbf{B}$ must be used to provide the correction. QED.

External Mapping Correction

An alternative approach for systems that do not have built-in mapping correction is to apply $\mathbf{A}$ to the command in the user's computer before issuing it to the controller and apply $\mathbf{B}$ to the read-back before displaying it.

DYNAMIC PERFORMANCE

So far we have considered the performance of open and closed-loop controllers with non-varying command inputs. In practice it is necessary to know how fast the stage will move and how well it will follow a varying command. Three properties govern the ability to follow varying commands:

1. the stiffness and mass of the stage

2. the frequency response of the controller

3. the current output capability of the piezo drive amplifier.

Open Loop Response

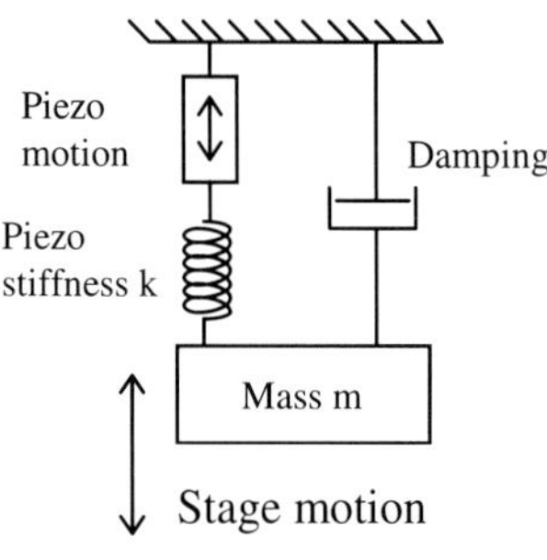

Fig. 3.7. Mass on a spring

The open loop response of a stage is generally dominated by the stage resonant frequency, which depends on the mass of the moving part and the stiffness of the support and piezo. This is a simple 'mass on a spring' as illustrated schematically in Fig. 3.7. The piezo stiffness is represented by a coil spring and mechanical damping by the dash-pot. This has a transfer function

$$G(s) = \frac{1}{\frac{1}{\omega_n^{\,2}} s^2 + \frac{1}{\omega_n Q} s + 1} \qquad (3.17)$$

where ω_n is the resonant frequency in rad·s⁻¹ ($\omega_n = 2\pi f_n$ where f_n is the resonant frequency in hertz) and Q is the amplification factor. The resonant frequency depends on the stiffness, k, of the stage and the mass, m, of the moving part and load

$$f_n = \frac{1}{2\pi} \sqrt{\frac{k}{m}} \qquad (3.18).$$

The Q of the system is the ratio of the response at resonance to the response at dc. It also governs the decay time of the ringing caused by a step input. Q depends on the losses in the system or 'damping': the higher the damping the lower the Q value.

Fig.3.8 shows this simple resonant open-loop transfer function with plots of its frequency and phase response, and time response to a 1 µm command step input. The example shown has a resonant frequency of 1 kHz and Q of 20. The plot shows the frequency response in decibels (dB): a factor of 20 is 26 dB.

> *The bel is a unit named after Alexander Graham Bell (1847-1922), the inventor of the telephone. It is used to measure ratios of intensities or powers and is defined as $log_{10}(P_1/P_2)$ and is famous because it is never used. Instead we have the decibel which is one tenth of a bel and is thus: $dB=10log_{10}(P_1/P_2)$. More commonly however the decibel is used to measure ratios of amplitudes, be they voltage, current, displacement etc.. Power is proportional to the square of amplitude, so the definition now becomes: $dB=10log_{10}(A_1^2/A_2^2)$ or: $dB=20log_{10}(A_1/A_2)$.*

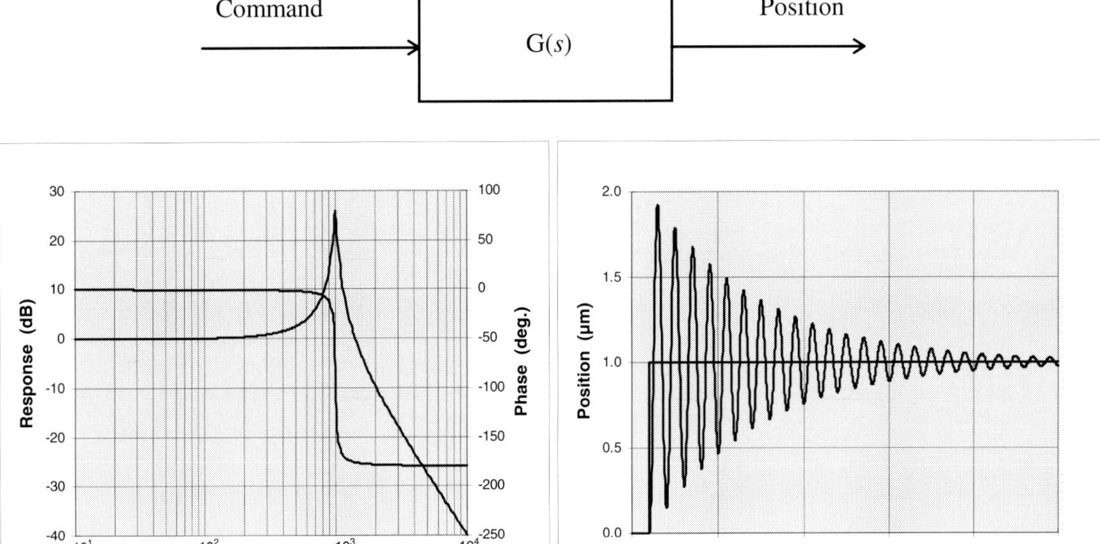

Fig.3.8. Open loop resonant stage

Closed Loop Models

In the previous discussion on static response we showed that an integrator in the forward loop (part of G) reduced the static error to zero. We also assumed that the feedback gain, H, was unity. What happens when we close the loop including the resonant response of Fig. 3.8?

Fig. 3.9. Closed-loop Control

Fig. 3.9 shows a block diagram with integrator and feedback. Things have suddenly got a bit complicated. As well as the integrator we have

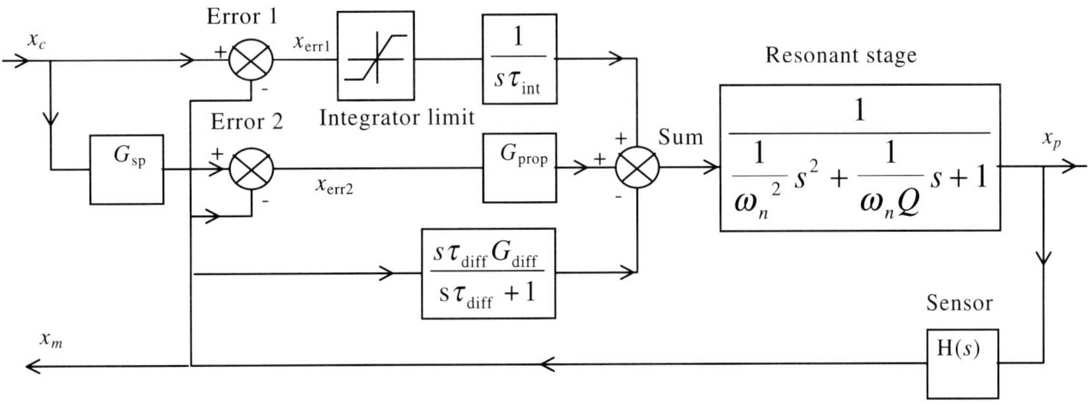

now added differential and proportional forward paths to create something like a classic Proportional, Integral and Differential (PID) controller. As will be shown, this can improve the performance enormously.

Integral Path

The integral path goes along the top of the diagram. Starting at the top left, x_c is the command input, in this case a step waveform, which is compared with the sensor output by the block 'Error 1'. The error signal output, x_{err1}, can be clipped at a settable level before being integrated, the reason for doing this will be shown later. The integrated error signal is then added to the proportional and differential terms by the 'Sum' block and passed to the resonant stage. The 'Sensor (H(s))' block measures the resultant stage motion to give the measured position x_m.

Proportional Path

The next layer down generates the proportional term. In a classic PID controller the error signal used for the proportional term would be the same as that used for the integral term. However the response can often be improved by varying the amount of command signal used to generate the proportional term. This is known as 'set point weighting' and is controlled by the gain block 'G_{sp}', the set point gain.

Differential Path

This departs from the classic PID implementation which again would act on the same error signal as the integral term. In general it is best to differentiate only the position information. The differential signal then gives a term proportional to the velocity of the moving stage and thus provides classic velocity feedback without the aid of a tachometer. Velocity feedback can greatly help in damping out resonances .

The transfer function used in the differentiator is not a pure differential term (which would be $s\tau_{diff}$) as the latter has very high gain at high frequency. This can introduce excess noise into the system with no real advantage in respect of resonance control. The transfer function used rises to unity at high frequency so the form of the function is set by τ_{diff} and G_{diff}. G_{diff} is the effective maximum gain along this path.

Resonant Stage

The stage is a simple resonant structure as used in the open loop model.

Response Curves and Settling Time

The frequency and step response depends on the amplitude of the command. With small signals the response depends mainly on the PID parameters chosen, the large signal response depends mainly on the slew-rate. The various domains are described in the next two sub-sections.

Small signal response

Small signal response depends almost entirely on the PID parameters and is characterised by a frequency v. amplitude curve and response to a step input

Settling time is defined as the time taken to get to within a certain percentage of the command step, we generally use 2 %. Fig.3.10 shows the frequency and step response for the control loop with just the integrator operating. The table gives the loop parameters and the resultant settling time.

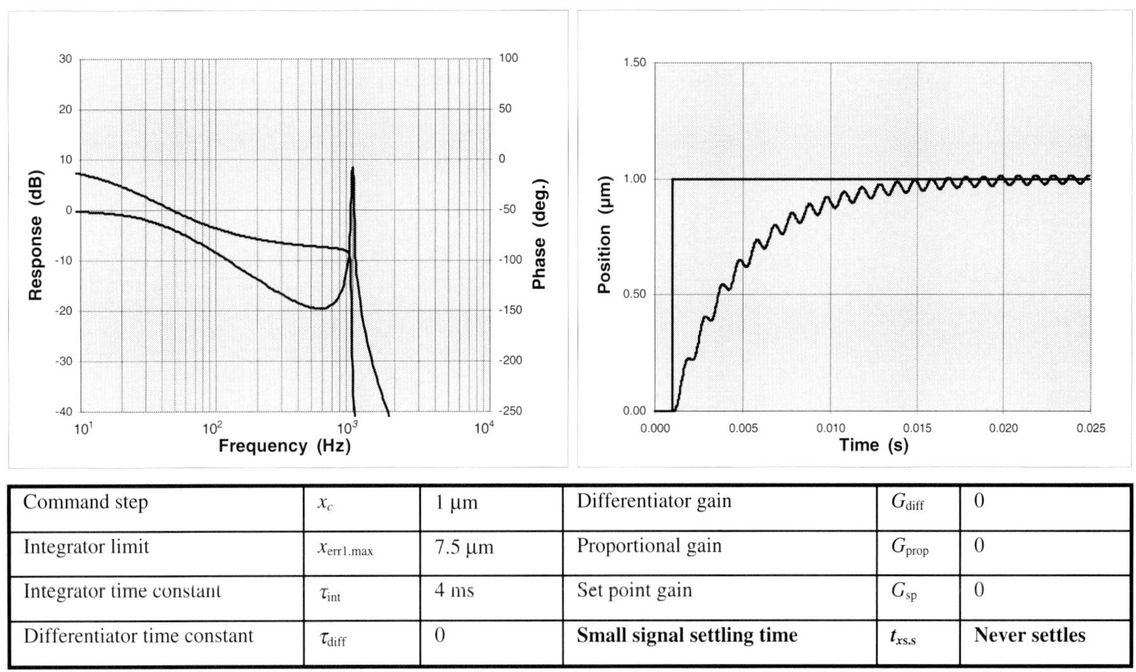

Command step	x_c	1 μm	Differentiator gain	G_{diff}	0
Integrator limit	$x_{err1.max}$	7.5 μm	Proportional gain	G_{prop}	0
Integrator time constant	τ_{int}	4 ms	Set point gain	G_{sp}	0
Differentiator time constant	τ_{diff}	0	**Small signal settling time**	$t_{xs.s}$	**Never settles**

Fig. 3.10. Integrator only

The response plotted is the fastest that can be done with just the integrator operating, reducing the integrator time constant results in uncontrollable oscillation. As can be seen the step response is not very fast and ringing is permanently present from the resonant stage. The frequency response also is very poor, starting to roll off at just a few Hz and the peak at the stage resonant frequency is still clearly visible.

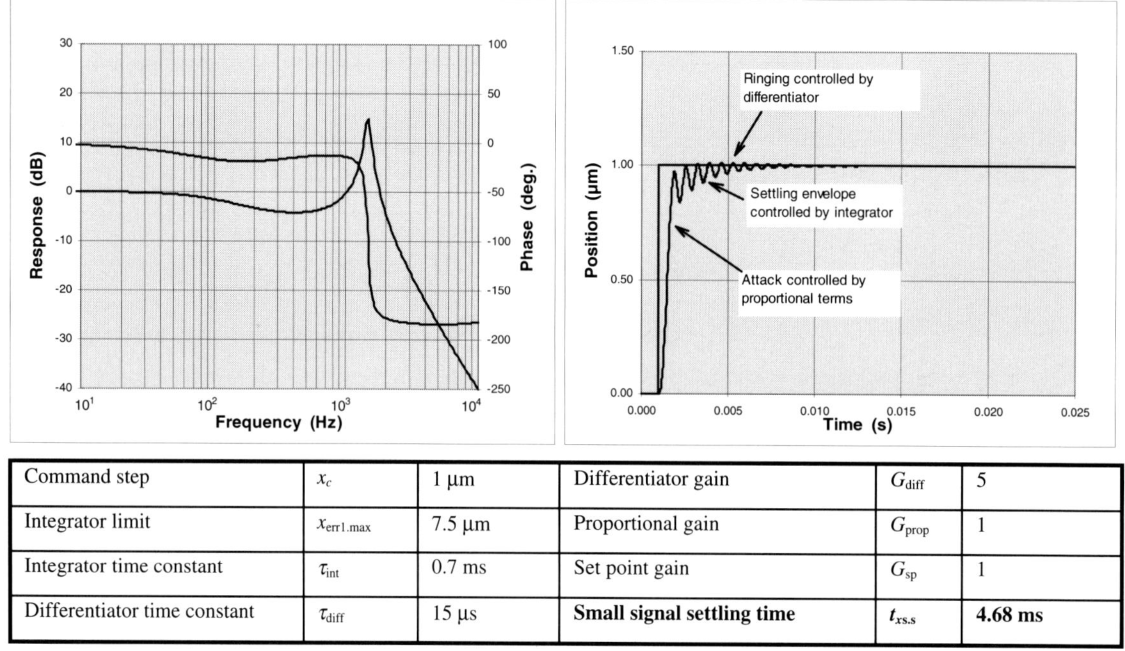

Command step	x_c	1 μm	Differentiator gain	G_{diff}	5
Integrator limit	$x_{err1.max}$	7.5 μm	Proportional gain	G_{prop}	1
Integrator time constant	τ_{int}	0.7 ms	Set point gain	G_{sp}	1
Differentiator time constant	τ_{diff}	15 μs	**Small signal settling time**	$t_{xs.s}$	**4.68 ms**

Fig .3.11. Proportional, integral and Differential

Fig. 3.11 shows a somewhat better response and illustrates the effect of the various loop parameters. In general the differential terms combat ringing (they add damping), the integrator controls the overall settling envelope and the proportional terms affect the attack at the start of the step. In this example the frequency response is a bit flatter and the settling times have improved, but the performance is not yet as good as it can be.

Fig. 3.12 shows a response optimised for settling time, now things are a lot better. The oscillations have been damped out completely and the frequency response is flat and well behaved up to a few hundred hertz.

This is good but what happens when a large command step is applied?

Large signal response

Fig. 3.13 shows the response to 50 Hz sine and square waves ranging in amplitude from 1 μm to 6 μm. The system parameters are as for Fig. 3.12. As can be seen, the response to steps greater than about 1 μm or sine waves greater that about 3 μm peak is positively ghastly.

The problem comes from the finite drive capability of the piezo drive amplifiers. This limits how fast the stage can move, and for large signals the stage will be asked to 'slew' at this maximum rate. The slew rate due to the drive amplifiers in the system described is 1 μm·ms^{-1}. While the drives are slewing, the servo loop is effectively out of control.

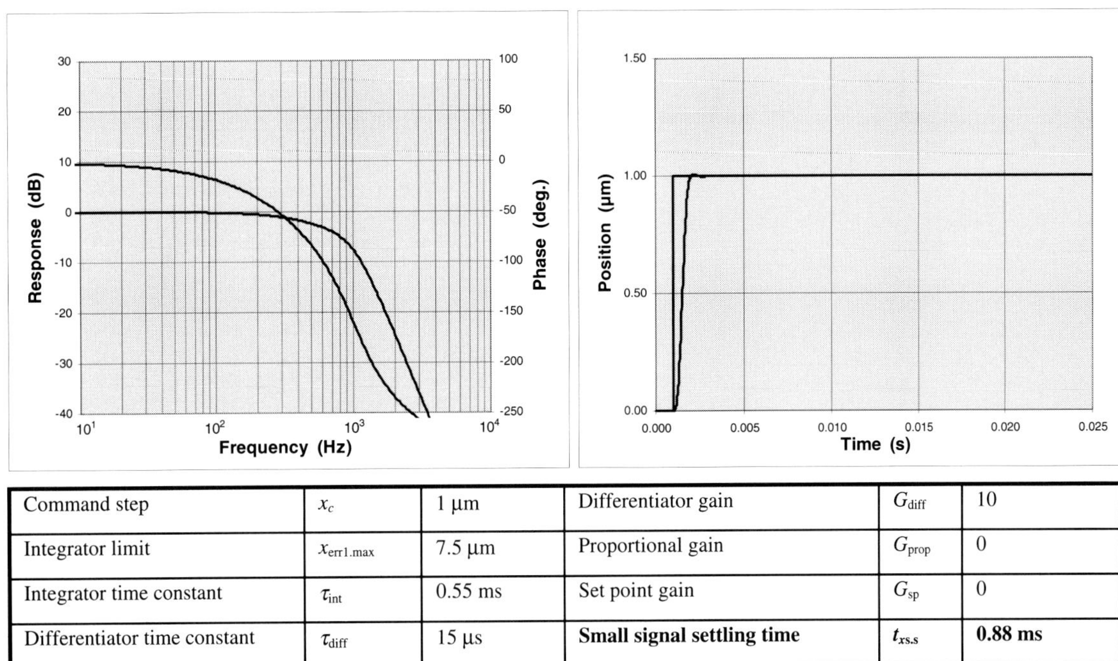

Command step	x_c	1 µm	Differentiator gain	G_{diff}	10
Integrator limit	$x_{\text{err1.max}}$	7.5 µm	Proportional gain	G_{prop}	0
Integrator time constant	τ_{int}	0.55 ms	Set point gain	G_{sp}	0
Differentiator time constant	τ_{diff}	15 µs	**Small signal settling time**	$t_{xs.s}$	**0.88 ms**

Fig. 3.12. Optimum Response

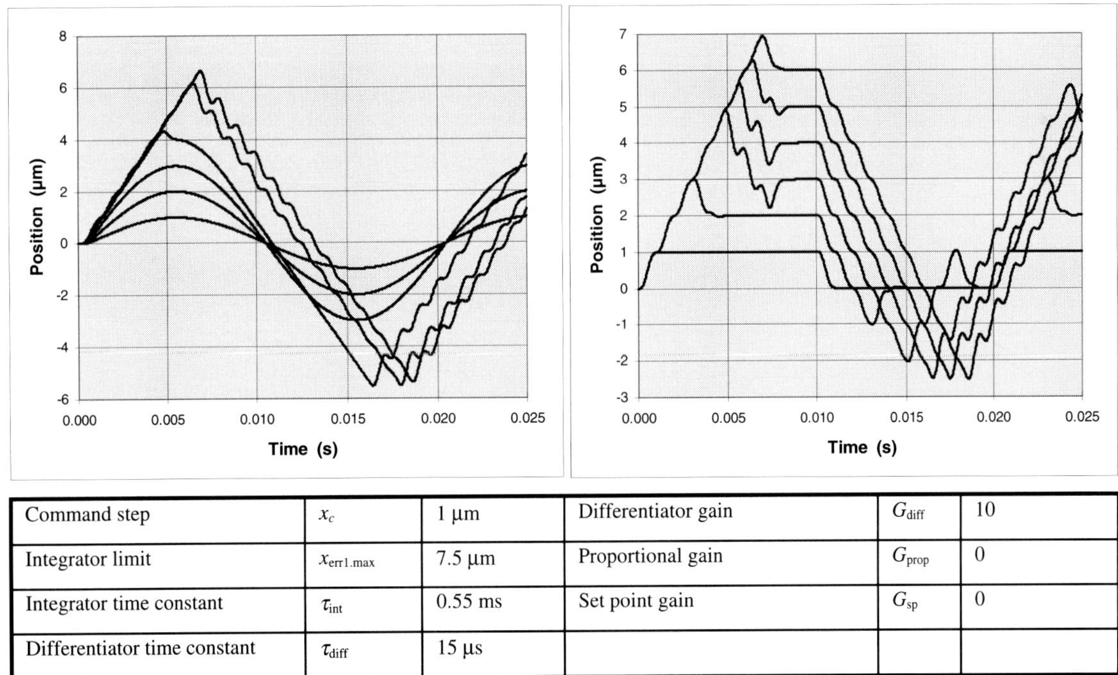

Command step	x_c	1 µm	Differentiator gain	G_{diff}	10
Integrator limit	$x_{\text{err1.max}}$	7.5 µm	Proportional gain	G_{prop}	0
Integrator time constant	τ_{int}	0.55 ms	Set point gain	G_{sp}	0
Differentiator time constant	τ_{diff}	15 µs			

Fig. 3.13. Large signal response

and the integrator will 'wind up' trying to get the piezos to move where they should be faster than they can. This causes the overshoot and, in some cases, completely chaotic response.

The answer to this problem is to limit the slew rate of the integrator to be equal to or less than that of the drive amplifiers. This is done by limiting the amplitude of the signal going into the integrator. If there is a constant (clipped) signal applied to the integrator its output will ramp at a rate given by

$$u_{\text{int.max}} = \frac{x_{\text{err1.max}}}{\tau_{\text{int}}} \tag{3.19}$$

where τ_{int} is the integrator time constant and $x_{\text{err1.max}}$ the integrator input limit. The integrator time constant used is 0.55 ms so limiting the input to 0.55 µm gives an integrator slew rate of 1 µm·ms^{-1}, the same as the piezo drive. The result of doing this is shown in Fig. 3.14. This is a far better response and in Queensgate's digital systems $x_{\text{err1.max}}$ is settable by the user to optimise the large signal performance.

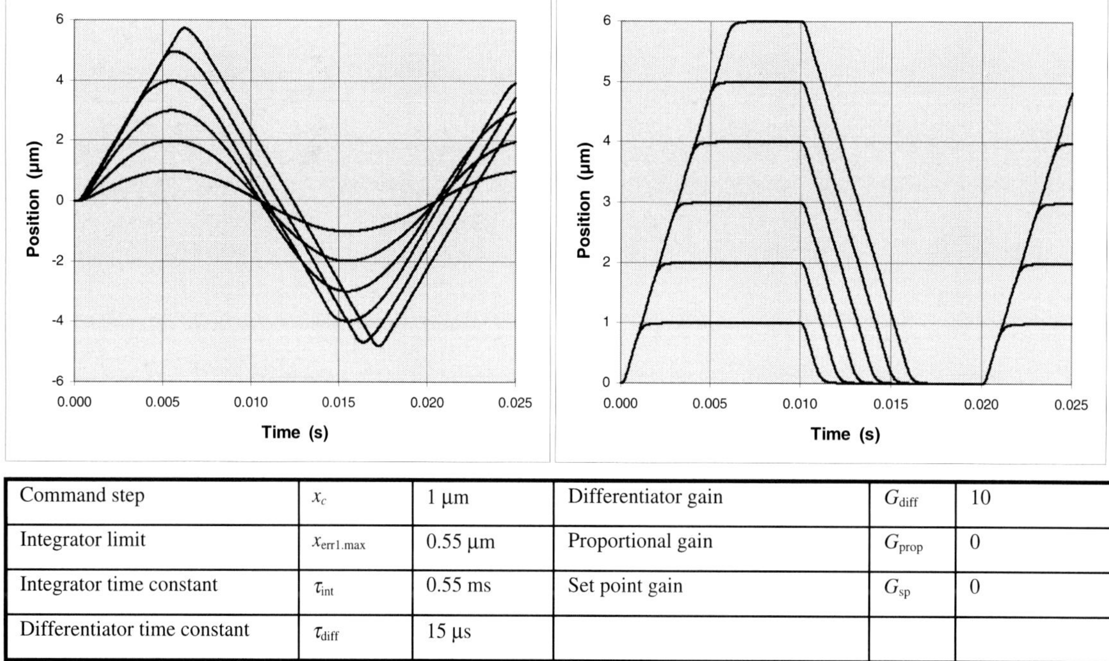

Command step	x_c	1 µm	Differentiator gain	G_{diff}	10
Integrator limit	$x_{\text{err1.max}}$	0.55 µm	Proportional gain	G_{prop}	0
Integrator time constant	τ_{int}	0.55 ms	Set point gain	G_{sp}	0
Differentiator time constant	τ_{diff}	15 µs			

Fig. 3.14 Optimised large signal response

What are small and large signals?

This depends on the nature of the command signal and the system in question. We can now define in general terms that a large signal is any signal that causes the stage to slew. For a step input, this is any command larger than the integrator input limit, for a sine command it depends on the sine amplitude and frequency. For a sine input, the maximum rate occurs at the zero crossing point and is given by

$$u_{sine.max} = 2\pi f A \tag{3.20}$$

where f is the frequency and A the amplitude. Thus for a 50 Hz sine input, the maximum amplitude that can be followed without slewing is 3.18 µm. This can be seen in the Fig.3.14, the 3 µm sine wave is just not slewing at the zero crossing point but the 4 µm one definitely is.

The converse of equation 3.20 is that any sine signal for which $Af <= u_{sine.max}/2\pi$ is a small signal. In this case the frequency response is given by the frequency response curve, such as that in Fig.3.12.

Large signal settling time

Obviously this depends on the slew rate and the amplitude of the command signal step. In general it is given by

$$t_{xs.l} = \frac{A}{u_{int.max}} + t_{xs.s} \tag{3.21}$$

where $t_{xs.s}$ is the small signal settling time.

SUMMARY

Stage positions can be controlled much more accurately using a closed-loop controller. The static and dynamic performance of the controller is determined by its PID parameters and integrator slew rate, all of which are user settable in Queensgate's digital control systems. This allows inherently resonant systems to be controlled with a precision far greater than could be achieved with open loop or simple analogue closed loop controllers.

APPENDIX

Reversing power series

Given a power series

$$y = ax + bx^2 + cx^3 + dx^4 + ex^5 + fx^6 + \ldots\ldots \tag{3.22}$$

the coefficients of the reverse series

$$x = Ay + By^2 + Cy^3 + Dy^4 + Ey^5 + Fy^6 + \ldots\ldots \tag{3.23}$$

are given by

$$A = \frac{1}{a} \tag{3.24}$$

$$B = -\frac{b}{a^3} \tag{3.25}$$

$$C = \frac{1}{a^5}\left(2b^2 - ac\right) \tag{3.26}$$

$$D = \frac{1}{a^7}\left(5abc - a^2d - 5b^3\right) \tag{3.27}$$

$$E = \frac{1}{a^9}\left(6a^2bd + 3a^2c^2 + 14b^4 - a^3e - 21ab^2c\right) \tag{3.28}$$

$$F = \frac{1}{a^{11}}\left(\begin{array}{l} 7a^3be + 7a^3cd + 84ab^3c - a^4f - 28a^2b^2d \\ -28a^2bc^2 - 42b^5 \end{array}\right) \tag{3.29}.$$

Derivation of reverse coefficients for higher order series is left as an exercise for the reader! Alternatively see Dwight 1961.

These formulae can be used to derive the 'b' coefficients if the 'a' coefficients are known, or vice versa.

REFERENCES

H. B. Dwight, Tables of Integrals and other Mathematical Data, *MacMillan Publishing Co., Inc. New York 1961*

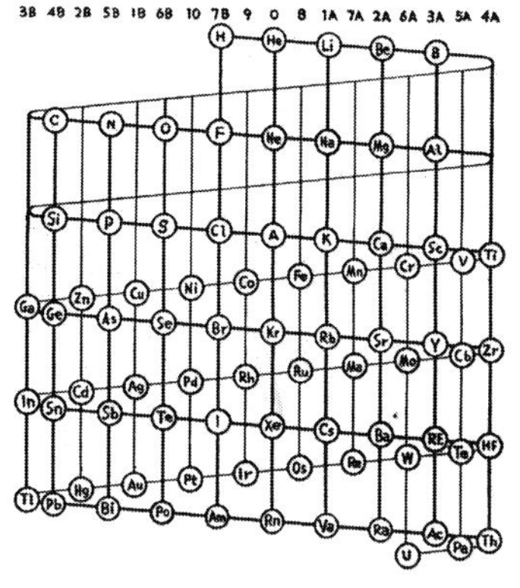

General considerations of material selection in conventional mechanical designs remain relevant in the design of precision instruments, however the dominant concerns may be different: for example strength and mass may not matter too much but the ability to maintain stability of shape and dimension, often to a high degree, probably does. Because material is used in small amounts, material cost may not have a significant influence on the total cost. Therefore higher priority can be given to obtaining the limits of possible performance and use of a wide range of exotic materials may become feasible.

In this Chapter:

DIMENSIONAL STABILITY

Thermal properties

Thermal properties of construction materials are always a major concern for both designing and using precision instruments. In normal use, all mechanical devices encounter heat inputs caused by environmental temperature change, by power dissipation in actuators, by operator handling and so on. The direct effect of the thermal disturbance is the thermal expansion which will cause dimensional changes in mechanical components, resulting in the loss of instrument accuracy. The dimensional change of a material due to a change in temperature is characterised by its Coefficient of Thermal Expansion (CTE), which varies tremendously with different materials. The CTEs of some commonly used materials in precision engineering can be found in Table 4.1. For example, typical values are $23 \times 10^{-6} \cdot K^{-1}$ for

Parameter	Aluminium 2024	Aluminium 7075	Stainless steel	Invar 36	Super-invar	Zerodur	Units
CTE	22.9	23.4	12	1.26	0.31	0 ± 0.05	$10^{-6} \cdot K^{-1}$
Thermal conductivity	119-190	142-176	25	11.1	10.4	1.64	$W.m^{-1} \cdot K^{-1}$
Young's modulus	73.1	71.7	190-210	147	148	90.6	GPa
Yield strength	76-393	103-503	280-700	276-414	303	90 (maximum strength)	MPa
Density	2.77	2.79	7.85	8.05	8.14	2.53	$10^3 \, kg \cdot m^{-3}$

Table 4.1 Material properties

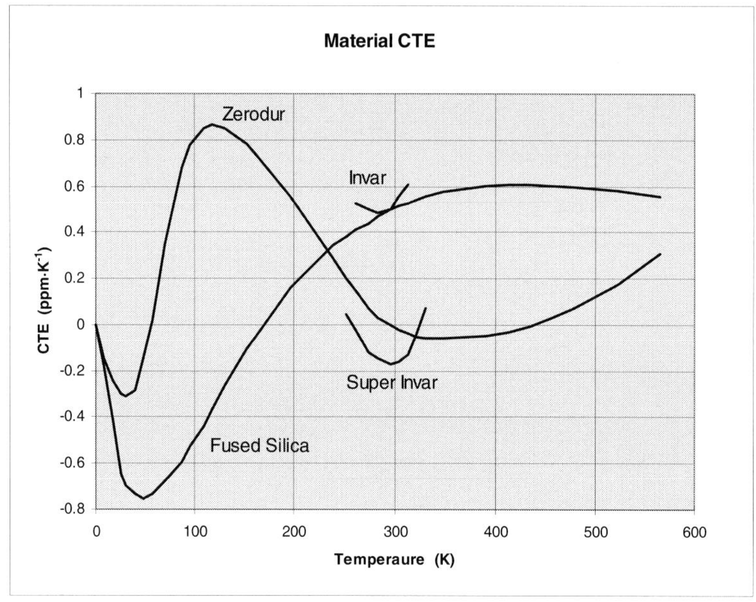

Fig. 4.1 CTE of various materials

aluminium alloys, and $0.3 \times 10^{-6} \cdot K^{-1}$ for SuperInvar. This means that 10 mm of material will expand 230 nm per degree for aluminium alloys and 3 nm per degree for SuperInvar. It is obvious that temperature stability can not be ignored when working in the nanometre regime.

It has to be noted that the CTEs of all the materials vary with temperature as shown in Fig. 4.1 (from Berthold, et al, 1976). From Fig. 4.1 we can see that Superinvar and Zerodur have very low CTE at room temperature, while fused silica has zero thermal expansivity near 170 K. In general, to reduce the thermal effect, construction materials with minimal thermal expansion coefficient should be used. However in some cases low thermal expansivity is not as useful as is the close expansivity match between the device and its mounting. Moreover, corrections to cope with thermal expansion are possible through control methods: the temperature can be measured and used to provide a correction.

Another problem is thermal gradients. They cause structure distortion, for which compensation is not possible. To avoid the effects of thermal gradients, the materials can be chosen either with low thermal expansivity (the material does not respond to temperature changes) or with high conductivity (the system reaches thermal equilibrium quickly).

Temporal stability

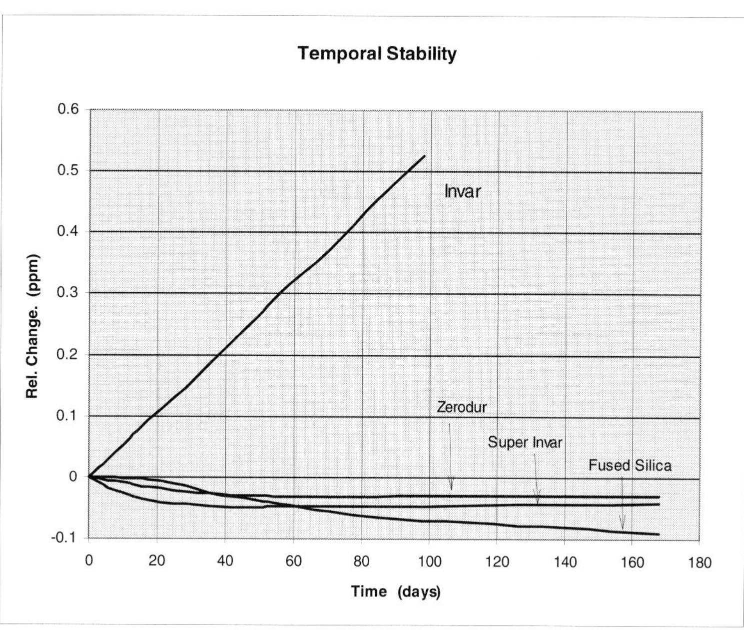

Fig. 4.2 Temporal Stability

Temporal stability refers to a material's dimensional changes with time without external force applied to the sample. If the thermal stability is considered as a short term effect, then the temporal stability can be taken as a long term effect. There is very little data on this property, except for some very low thermal expansion materials such as Invar, Superinvar and Zerodur etc.. These materials are normally used for construction of precision instruments and machine tools therefore their stability, both thermal and temporal, is of great concern. For example, Invar's low thermal expansion coefficient is about $1.26 \times 10^{-6} \cdot K^{-1}$ but its isothermal length change with time, however, can be as large as $0.1 \times 10^{-6} \cdot day^{-1}$. The elongation with time eventually halts, but this may take as long as 30 years (Physics and applications of Invar alloys, Honda Memorial Series on Materials, 1978). This period can be shortened by proper heat treatments. Delayed elastic deformation of Zerodur at room temperature upon release of external forces applied for a significant time interval has been observed by many researchers (Pepi and Golini, 1991; Yoder,1993). The temporal change of sample length for a number of low thermal expansion materials is shown in Fig. 4.2 (From Berthold et al., 1976). Indeed, with reasonable control of the ambient temperature, pressure, humidity and magnetic field, temporal instability may be the factor limiting performance.

MECHANICAL PROPERTIES

Generally speaking, mechanical design can do a lot to satisfy the needs for stiffness and strength, providing the space is not too highly restricted, so the value of Young's modulus and yield strength are less important than thermal properties for the design of precision instruments. However, to reduce the effects of the environment many precision devices are deliberately designed to be small. Then the mechanical properties of materials have to be carefully considered. For example, the material's strength may limit the maximum range which can be achieved by flexure mechanisms; a low Young's modulus may not be able to provide a sufficient stiffness for the NanoMechanism or its frame, which is sometimes used as the metrological datum; the hardness may affect the contact stiffness between the mechanism and its actuators, which has a direct effect on the resonant frequency of a mechanical system. Also the mass of material can make a big difference to the dynamic properties of NanoMechanisms. For example the density ratio of SuperInvar and aluminium alloys is about 3, so the resonant frequency of an aluminium system can be $\sqrt{3}$ times higher than that of a SuperInvar system if the stiffness of the systems are the same. In the design of flexure hinge mechanisms, it is important that the materials have good fatigue properties so that the lifetime is maximised. Normally the fatigue strength is defined as the maximum stress under which the samples will not fail after 10^7 stress cycles for steels and 5×10^8 for aluminium. The fatigue life of components are affected by many factors including material, geometry, surface roughness, machining method, heat treatment, surface coating and so on (Collins, 1993).

Machineability

Machineability of material is another limit for the design of NanoMechanisms. First of all, the material selected has to be machineable for required geometrical figures. For example, most of our flexure stages are cut with electro-discharge machining. Glass ceramics are obviously not available for this application although they have many good properties. On the other hand, machining cost dominates the price of the products because most of the components in NanoMechanisms are relatively small in size, so the effect of material cost is not significant. Machineability of material depends on material properties such as strength, hardness, toughness and thermal conductivity, etc..

SOME TYPICAL MATERIALS

Zerodur

Zerodur is a transparent, homogenous glass ceramic with an exceedingly low coefficient of thermal expansion compared with other low expansion materials: about $0.1 \times 10^{-6} \cdot K^{-1}$ at temperatures between -30 and +70 C. Its brittleness restricts its application in systems carrying tensile and bending loads. It is normally manufactured by grinding with diamond tools, therefore can only be used for components of very simple shape. Zerodur is a good substrate material for capacitance sensor pads because of its ideal dielectric properties, long time stability and chemical stability.

Invar and SuperInvar

Invar and SuperInvar are alloys of iron and nickel with very low thermal expansivity. Invar contains 36 % nickel with small quantities of manganese, silicon and carbon, having a typical thermal expansion coefficient of $1.3 \times 10^{-6} \cdot K^{-1}$. Substitution of 5 % cobalt for 5 % nickel provides SuperInvar which has an even lower thermal expansion coefficient ($0.3 \times 10^{-6} \cdot K^{-1}$) and higher temporal stability than Invar. Their CTE remain stable and low over a limited range (typically 4 to 38 C for Invar) and the CTE changes relatively rapidly outside this range. Note that SuperInvar undergoes a phase change when cooled to below -50 C so it is not a good material of choice when exposure to such low temperatures is anticipated. For maximum stability and minimum CTE parts need to be properly heat treated. Invar and SuperInvar can be fabricated using conventional metalworking techniques and so are easy to use for components of complex shapes. Their toughness and elasticity make them suitable for flexure mechanisms although high density may be a disadvantage for design if high resonant frequency is required. The fact that they are magnetic can be used to advantage in low strain mounting schemes.

Aluminium alloy

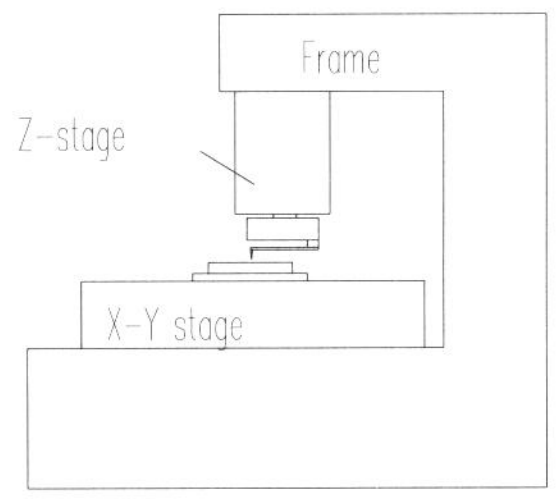

Fig. 4.3 Atomic Force Microscope

Aluminium alloy is one of the most commonly used materials in engineering construction. Precision instruments mainly use its properties of good thermal conductivity, ease of manufacturing (low machining cost) and low mass. It has to be carefully used because of its high thermal expansion. This material is often chosen for thermal matching. Taking the Atomic Force Microscope (AFM) in Fig.4.3 as an example, the X-Y scanning stage forms a part of metrological loop. If the material of the frame, on which both probe and scanning stage are mounted, is aluminium, then the stage has to be made of same material to compensate the thermal expansion error from the frame.

Besides, when low mass or high resonant frequency are required aluminium becomes a very attractive material due to its low density.

Stainless steel

Some precision instruments are built of stainless steel. The thermal expansion coefficient of stainless steel is lower than aluminium, but its thermal conductivity is poor. Therefore, the thermal effects are still a big problem. Its machinability is similar to SuperInvar.

Rusting resistance is the main merit of using the material. No plating or other surface treatments are needed, which would tend to increase the fatigue life of the components. Its high Young's modulus makes stainless steel more advantageous than aluminium when high stiffness is needed.

REFERENCES

Berthold J.W. III, Jacobs S.F., and Norton M.A., Dimension stability of fused silica, Invar, and several ultralow thermal expansion materials, *Appl. Opt., 8, 1898, 1976*

Collins J.A, Failure of materials in Mechanical design, *John Wiley and Sons, Inc. 1993*

Pepi J. and Golini D, "Delayed Elastic Effects in the Glass Ceramics Zerodur and ULE at Room Temperature, *Appl. Opt., 30, 3087, 1991*

Physics and applications of Invar alloys, Honda Memorial Series on Materials, *Science 3, Maruzen, Tokyo, 1978*

Yoder P.R, Jr., Opto-mechanical systems design, *Marcel Dekker, Inc., New York, 1993.*

CHAPTER 5. CAPACITANCE SENSORS

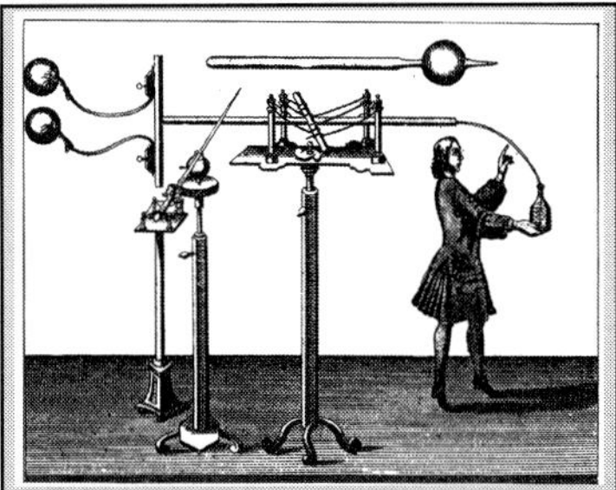

The Leyden jar used water in a jar as one terminal of a capacitor, connection was made by a chain dangling into it. The other terminal of the capacitor was the worker's hand holding the jar. It would appear that the amount of charge stored was measured by the strength of the electric shock received when the chain was removed.

In later versions the hand was replaced by a layer of tin foil.

Michael Faraday (1791 to 1867) after whom the unit of capacitance, the farad, is named is actually remembered more for his work on electromagnetic phenomena than capacitance. He did, however do a lot of work on the relative permittivity of various materials

The concept of electrical capacitance was first reported by Pieter van Musschenbroek at the University of Leyden in 1745 while trying to store static electricity in a jar of water (the Leyden jar). However one has to wait until 1873 for Maxwell to give a concise definition of capacitance.

In this chapter:

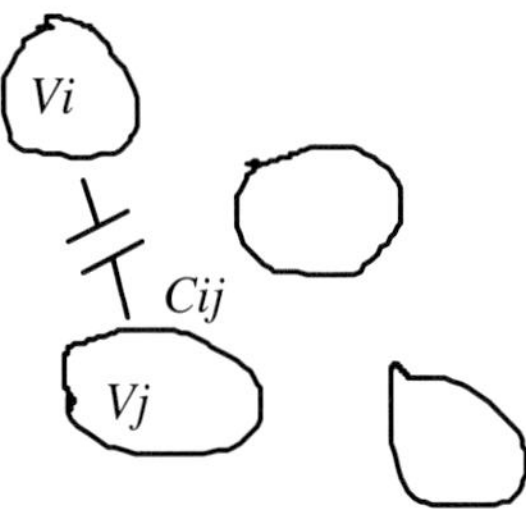

Fig. 5.1. Isolated conductors

Capacitance relates to the ability of electrically isolated conductors to store charge. Given an arbitrary arrangement of conductors as illustrated in Fig. 5.1, the capacitance C_{ij} between conductors i, j is given by

$$C_{ij} = \frac{Q_{ij}}{V_i - V_j} \tag{5.1}$$

where Q_{ij} is the charge on conductor i induced by the potential difference V_i-V_j; V_i is the potential on conductor i and V_j is the potential on conductor j (Maxwell, 1873; reviewed more recently by Heerens, 1986). It should be noted that the capacitance between conductors i and j relates only to *their* potentials V_i, V_j and not to the potential of adjacent conductors. The *presence* of adjacent conductors, not their potential, influences the capacitance.

LONG AND SHORT-RANGE MICROMETERS

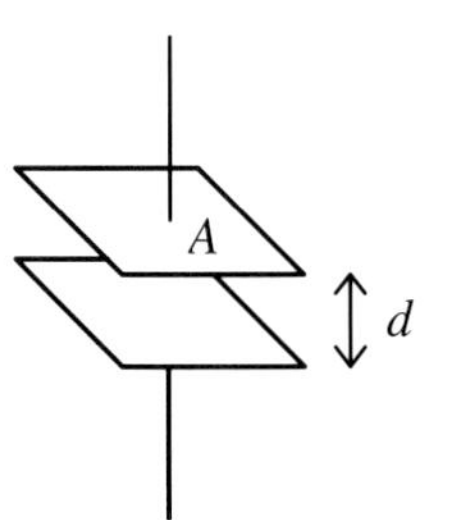

Fig. 5.2. A parallel plate capacitor

The simplest form of practical capacitor is the parallel plate capacitor shown in Fig. 5.2. The capacitance, C, is given by

$$C = \frac{\varepsilon_r \varepsilon_0 A}{d} \tag{5.2}$$

where ε_r is the relative permittivity of the medium between the plates; ε_0 is the permittivity of vacuum; A is the plate area and d the plate separation or gap. As all the text books say, this equation ignores edge effects. More of this later.

Given that one can measure capacitance, the parallel plate capacitor can be made into a micrometer by letting the displacement to be measured vary the gap, d, or the area, A. Area can be varied by letting the displacement vary the overlap between the plates, for example as shown in Fig. 5.3a. A variation of this shown in Fig. 5.3b is the familiar rotary tuning capacitor that can be used to measure rotation. These configurations are fine for measuring large displacements, but care must be taken to ensure there is little gap change as a small change in gap can give a large change in capacitance. This can be minimised (but not eliminated) by using coaxial cylinders for the linear case and a multi-vaned arrangement for the rotational case.

The 1986 recommended values of the physical constants defines ε_0 as $1/\mu_0 c^2$ where μ_0 is the permeability of vacuum and c is the speed of light. The speed of light is defined as exactly 299 792 458 $m \cdot s^{-1}$ and the permeability of vacuum as exactly $4\pi x\ 10^{-7}\ N \cdot A^{-2}$ (Defining something as exactly related to an irrational number is an interesting concept!) To give some values:

c = 299 792 458 $m \cdot s^{-1}$
μ_0 = 12.566 370 614...x 10^{-7} $N \cdot A^{2}$
ε_0 = 8.854 187 817... $pF \cdot m^{-1}$

It is interesting to note that if the plates of a parallel plate capacitor are circular, then the pi's in the expression for the plate area and definition of ε_0 cancel and the capacitance is a rational number!

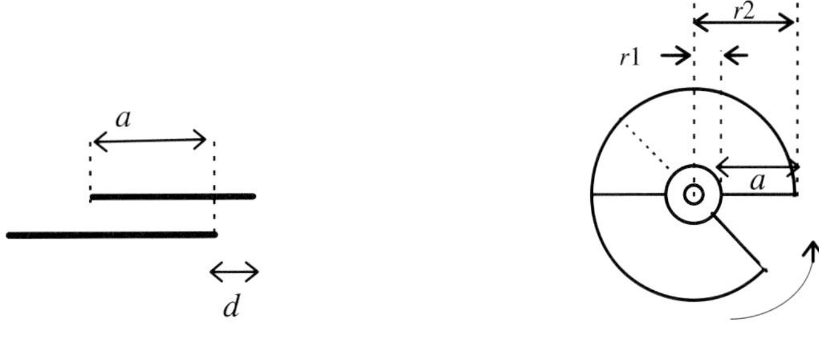

a). Linear micrometer   b). Rotational micrometer

Fig. 5.3. Long range micrometers: area change

Target and probe electrodes

The sensitivity to gap change does, of course, give us a more sensitive way of measuring displacement: let the displacement vary the gap. Fig. 5.4 shows implementations for measuring small linear displacements and rotations. One plate is made larger than the other to minimise the effects of motions at right angles to the displacement being measured: this really does work! The smaller plate is usually called the 'probe' and the larger one the 'target'.

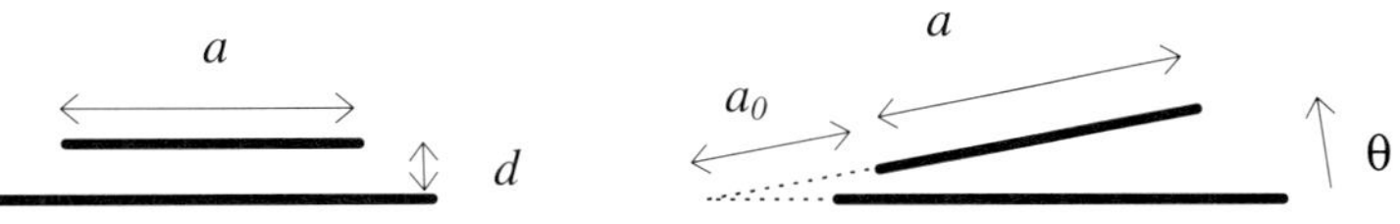

Fig. 5.4. Short range micrometers: gap change

a). Linear micrometer   b). Rotational micrometer

Sensitivities

The sensitivity of each configuration is approximately the same when considered in terms of fraction of the mechanical parameter being varied. For example if the effective plate area is doubled by changing the overlap, the capacitance is doubled; if the gap is doubled the capacitance is halved etc., the dynamic range is approximately the same. It is, however, instructive to put in some numbers.

Consider the cases with plates having dimensions a,b. The 'b' dimension is up out of the page. In the long range rotational case of Fig. 5.3b, 'a' is the difference between the external and internal plate radii $r2$ and $r1$. Let the nominal d value be d_0 and the nominal rotation θ_0. For computation of numbers, let:

$a = b = 10.63$ mm
$a_0 = r1 = 2$ mm
$d_0 = 0.1$ mm
$\theta_0 = 20$ mrad (short range: gap change)
$\gamma_0 = 1.4$ rad (long range: area change).

These numbers are chosen to make the nominal capacitances of each configuration approximately 10 pF, for comparison purposes.

Long-range linear (area change)

$$C = \frac{\varepsilon_r \varepsilon_0 ba}{d_0} = \frac{8.85 \times 10.63^2 \times 10^{-6}}{100 \times 10^{-6}} = 10.0 \text{ pF} \tag{5.3}$$

$$\frac{dC}{da} = \frac{\varepsilon_r \varepsilon_0 b}{d_0} = 0.94 \text{ fF} \cdot \mu\text{m}^{-1} \tag{5.4}.$$

Long-range rotational (area change)

$$C = \frac{\varepsilon_r \varepsilon_0 \left(r_2^2 - r_1^2\right)\gamma_0}{2d_0} = \frac{8.85\left(0.01263^2 - 0.002^2\right) \times 1.453}{2 \times 100 \times 10^{-6}} = 10.0 \text{ pF} \tag{5.5}$$

$$\frac{dC}{d\gamma} = \frac{\varepsilon_r \varepsilon_0 \left(r_2^2 - r_1^2\right)}{2d_0} = 6.88 \text{ pF} \cdot \text{rad}^{-1} \tag{5.6}.$$

Short-range linear (gap change)

$$C = \frac{\varepsilon_r \varepsilon_0 ba}{d} = \frac{8.85 \times 10.63^2 \times 10^{-6}}{100 \times 10^{-6}} = 10.0 \text{ pF} \tag{5.7}$$

$$\frac{dC}{dd} = \frac{C}{d_0} = \frac{10.0}{100} = 100 \text{ fF} \cdot \mu\text{m}^{-1} \tag{5.8}.$$

Short-range rotational (gap change)

$$C = \frac{\varepsilon_r \varepsilon_0 b}{\theta} \ln\left(1 + \frac{a}{a_0}\right) = \frac{8.85 \times 0.01063}{0.01734} \ln\left(1 + \frac{10.63}{2}\right) = 10 \text{ pF} \tag{5.9}$$

$$\frac{dC}{d\theta} = \frac{C}{\theta_0} = \frac{10}{0.01734} = 576 \text{ pF} \cdot \text{rad}^{-1} \tag{5.10}.$$

The sensitivities of the short-range configurations are thus about 80 to 100 times greater than the long range, though of course the useful range is correspondingly less. In general the useful range is about equal to the nominal value of the gap (short range) or plate overlap (long range). The results for these examples is summarised in Table 5.1.

Configuration	Range	Nominal capacitance	Sensitivity
Long-range linear	10 mm	10 pF	0.94 fF·μm^{-1}
Long-range rotation	1.5 rad	10 pF	6.9 fF·mrad^{-1}
Short range linear	100 μm	10 pF	100 fF·μm^{-1}
Short-range rotation	17 mrad	10 pF	580 fF·mrad^{-1}

Table 5.1. Sensitivities

CAPACITANCE MEASUREMENT CONSIDERATIONS

It will have been noticed that the capacitance is directly proportional to the displacement being measured in the long-range case, but is inversely proportional in the short-range case. Does this mean that the short-range configurations are inherently non-linear? The quick answer is no.

Consider a source of alternating voltage applied across a capacitor C (Fig.5.5). An alternating current I will flow given by

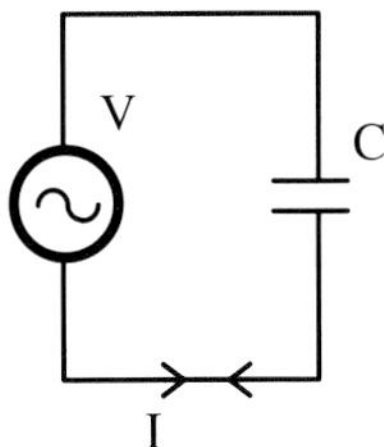

Fig. 5.5. Current through a capacitor

$$I = j\omega VC \tag{5.11}$$

where ω is the angular frequency of the alternating voltage, or $2\pi f$ where f is the frequency in hertz. The current is directly proportional to the capacitance, the j indicates that the current is 90° out of phase with the voltage. Thus if a constant amplitude voltage is applied to a long-range micrometer capacitor, the current will be directly proportional to the displacement or angle being measured. Current is relatively straightforward to measure so we have an inherently linear measuring device.

Now re-arrange equation 5.11 to give

$$V = \frac{I}{j\omega C} \tag{5.12}.$$

If a constant amplitude *current* is passed through the capacitor, the voltage developed across it is inversely proportional to the capacitance. Thus by applying a constant current the voltage will be directly proportional to displacement or angle in the short-range configuration. Now by measuring the voltage we have an inherently linear measuring device.

If, then, displacement or angle can be measured linearly, why did we spend so much time in Chapters 2 and 3 talking about mapping functions? Enter the real world.

PRACTICAL CAPACITORS

The capacitance of a parallel plate capacitor is usually calculated using equation 5.2, but as has been stated, this equation ignores edge effects. It also ignores the effect of plate tilt, flatness and finish, and equations 5.11 and 5.12 ignore the contribution of stray capacitance to C. All these things conspire to make sensors that depart from perfect linearity.

Queensgate is primarily concerned with the measurement and control of small displacements, so all our sensors work in the short-range gap change mode. Only the short range linear configuration will be considered from now on and with circular rather than rectangular plates.

The C of equation 5.2 can be modified to include the extra effects:

$$C' = C(1+\eta) + C_s \tag{5.13}.$$

C is also a function of electrode finish, flatness and tilt. η is a correction depending on the guard ring configuration, see below.

C_s is stray capacitance in parallel with the sensor capacitor and could arise from unshielded wiring or edge effects.

The trick now is to determine what effect these error terms have on the sensor performance.

Stray Capacitance

Ignoring η and tilt etc. for now, Equation 5.12 can re-written to include stray capacitance

$$V = \frac{I}{j\omega(C + C_s)} \tag{5.14}$$

which gives

$$V = \frac{Id}{j\omega(\varepsilon_r \varepsilon_0 A + C_s d)} \tag{5.15}.$$

This is obviously non-linear in d. Equation 5.15 is more conveniently written

$$V = \frac{Id}{j\omega(C_0 d_0 + C_s d)} \tag{5.16}$$

where C_0 is the sensor capacitor value at the nominal gap d_0.

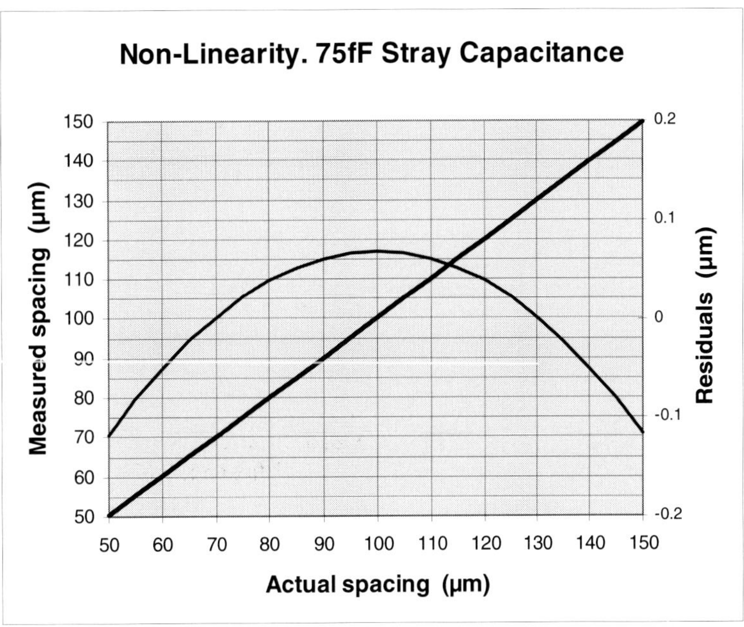

Fig. 5.6 Effect of stray capacitance

Fig. 5.6 shows a plot of equation 5.16 with C_0=10 pF; d_0 =100 µm and C_s=75 fF. I/ω has been normalised to give a nominal slope of unity, i.e. the measured spacing is nominally equal to the true spacing. The 'j' term just means that the voltage is 90° out of phase with the current and is ignored.

The residual curve (right hand scale) shows a maximum departure from linearity of about 0.1 µm in 100 µm or 0.1%. This is quite large considering that 75 fF is a relatively small capacitance and shows the importance of keeping stray capacitance in parallel with the sensor capacitor to a minimum. Of course if the sensor is part of a stage, the whole system can be calibrated and this effect taken out by means of the mapping function. However, as will be shown in the following sections, it can be designed out in the first place.

Guard rings

The simple short-range configuration of Fig. 5.4a suffers from edge effects that tend to make the capacitance larger than equation 5.2 would suggest. This deceptively simple configuration is one that has not (to the authors knowledge) been solved analytically using Laplace's equation, as the boundary conditions are hard to define (see Appendix). Jones and Richards (1973) consider the effect to be a constant stray capacitance, after the work of Scott and Curtis (1939) and quote a value that gives around 23 fF·mm[-1] of edge. Maxwell (1873, treatise 195 or thereabouts) says that the apparent area of a capacitor increases with gap, d, and is given by

$$A' = A + \frac{\ln 2}{\pi} ld \tag{5.17}$$

where l is the length of edge. As the area increases with gap, the capacitance increase is constant with gap and is

$$C_s = \varepsilon_r \varepsilon_0 \frac{\ln 2}{\pi} \tag{5.18}$$

per unit of edge. This works out as 2 fF·mm^{-1}, about ten times lower than quoted by Jones and Richards. To resolve the conflict, a capacitor with edge effects has been analysed at Queensgate using finite element analysis. Also the effect has been calculated to be approximately $\varepsilon_r\varepsilon_0/\pi\sqrt{2}$ or 2 fF·mm^{-1} by Hicks (1997) using integral techniques. We believe Maxwell! 2 fF·mm^{-1} seems about right.

As has been shown, stray capacitance whatever its cause is a bad thing. The example of Fig. 5.6 used a stray capacitance of 75 fF which is the extra contribution due to edge effects of our test capacitor with radius 6 mm. This edge-effect contribution can be minimised by using a guard ring and various guard configurations are shown in Fig. 5.7. They are:

a)   parallel plate capacitor with simple guard ring and thin plates

b)   parallel plate capacitor with simple guard ring and thick plates

c)   parallel plate capacitor with target surrounding guard

d)   parallel plate capacitor with guard surrounding target.

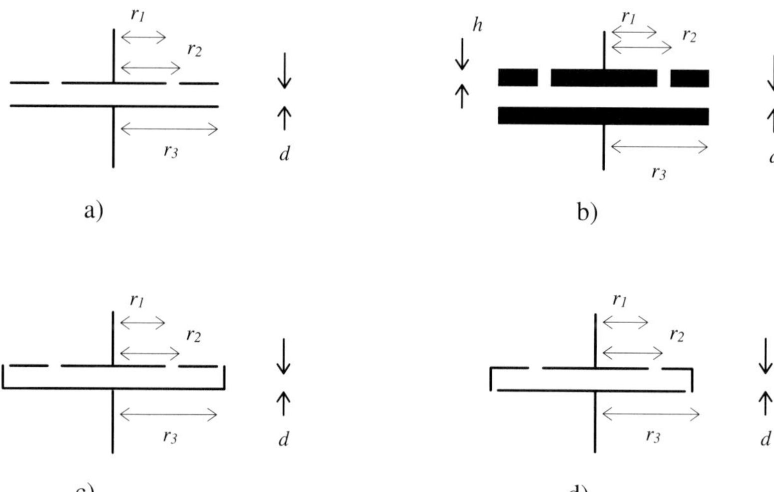

Fig. 5.7 Guard configurations

Configurations 'a' and 'b' are both in use in Queensgate products; 'c' and 'd' can be used for very precise systems, especially if radiated interference susceptibility is an issue.

First let us define some intermediate terms

$$r_i = \frac{r_1 + r_2}{2} \tag{5.19}$$

$$s = r_2 - r_1 \tag{5.20}.$$

Thin plates with simple guard ring

This configuration gives a capacitance very close to the 'perfect' case and is the one used most commonly by Queensgate, but there are errors due to the finite value of 's'. Heerens and Vermeulen (1975) quote a correction originally derived by Moon and Sparks (1948). In this case C in equation 5.13 is given by

$$C = \frac{\varepsilon_r \varepsilon_0 \pi r_i^{\,2}}{d} \tag{5.21}$$

and:

$$\eta = -\frac{s^2}{2\pi r_i d} \coth\!\left(\frac{\pi r_i}{d}\right) \tag{5.22}.$$

This is valid for

$$s << d \tag{5.23}$$

$$r_3 - r_2 > 5d \tag{5.24}.$$

The coth (hyperbolic co-tangent) term can usually be ignored.

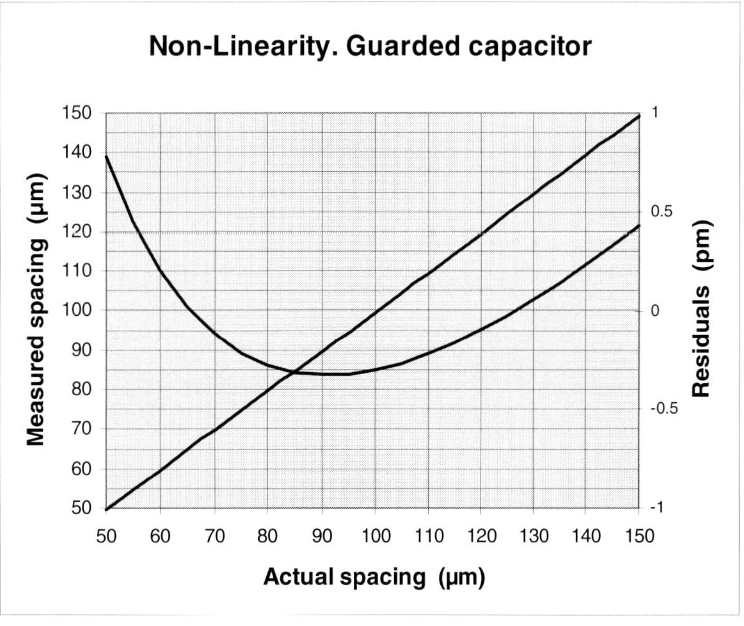

Fig. 5.8. Guarded capacitor

Fig. 5.8 shows a plot of measured spacing against true plate spacing incorporating the correction term of equation 5.22 but ignoring stray capacitance and other error terms. Again the slope is normalised to unity. The capacitor parameters are:

radius: 6 mm
guard gap: 25 µm

This gives a nominal capacitance of 10 pF at 100 µm electrode spacing.

As can be seen from the residual curve in Fig. 5.8, the maximum departure from linearity is less than 1 pm over the displacement range of 100 µm. This is a negligible linearity error showing that the guard ring really does reduce the edge effects. A simple guard ring is relatively easy to make and gives a significant improvement in linearity.

Other guard configurations

Error terms for the other configurations, 'b', 'c' and 'd' of Fig. 5.7, have been derived by various workers. Heerens and Vermeulen (1975) have used Laplace's equation to solve configurations 'c' and 'd' which are the only ones where the boundary conditions are well defined: the full answer is an infinite series of modified Bessel functions of the first and second kind with order zero, but there are valid approximations that are more manageable (Equations 5.27 and 5.28). Brown and Bullied quote a solution for 'b', thick capacitor plates, valid if h is five times d or greater, equation 5.25. This comes from the equations of Maxwell (1873): the factor of 0.22 is the ubiquitous $(\ln 2)/\pi$.

Configuration 'b'

$$\eta = \frac{s\left(1 + \dfrac{s}{2r_1}\right)}{r_1\left(1 + \dfrac{0.22s}{d}\right)} \tag{5.25}.$$

Equation 5.25 is valid for:

$$h \geq 5s \tag{5.26}.$$

Configuration 'c'

$$\eta = \frac{4d}{\pi r_i}\left(\frac{r_3}{r_i}\right)^{\frac{1}{2}} e^{-\frac{\pi(r_3 - r_i)}{d}} \tag{5.27}.$$

Configuration 'd'

$$\eta = -\frac{4d}{\pi r_i}\left(\frac{r_3}{r_i}\right)^{\frac{1}{2}} e^{-\frac{\pi(r_3-r_i)}{d}}$$ (5.28).

It is left as an exercise to plot out the non-linearity that would occur with these configurations!

Plate Tilt

Tilt of one capacitor plate with respect to the other can seriously affect linearity. Again if the sensors are mounted in a stage, this will be calibrated out, but care must be taken if the sensors are used stand-alone. There are many expressions for the variation of capacitance with plate tilt, see the Appendix to this chapter. A convenient one is derived by Harb *et al.* (1995)

$$C = \frac{2\varepsilon_r\varepsilon_0\pi r_i^2}{d}\left(\frac{1}{1+\sqrt{1-k^2}}\right)$$ (5.29)

where

$$k = \frac{r_i\sin 2\theta}{2d}$$ (5.30).

Fig. 5.9 shows capacitance as a function of tilt angle at 100 µm nominal gap for the test capacitor with 6 mm radius. The nominal gap is defined at the centre of the electrode. As can be seen, a 5 mrad tilt

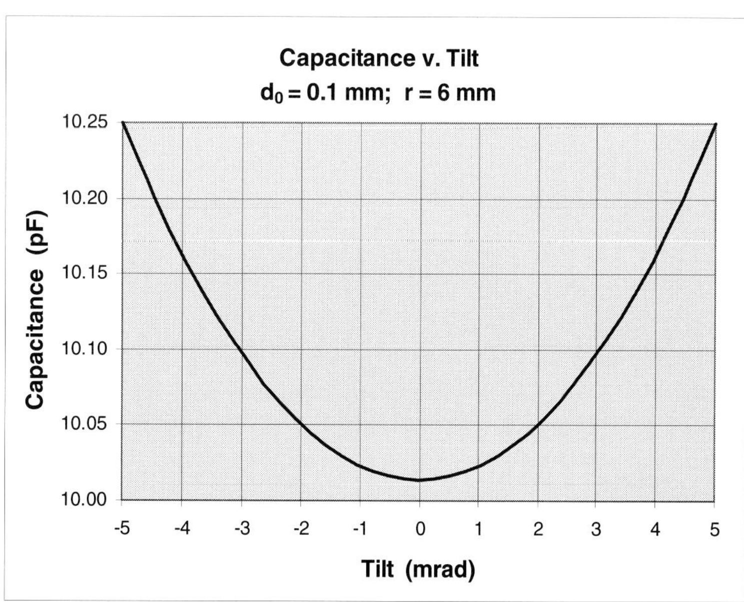

Fig. 5.9. Capacitance v. tilt
r=6 mm; d_0=100 µm; C=10 pF

changes the capacitance by 0.25 pF, which is quite a lot.

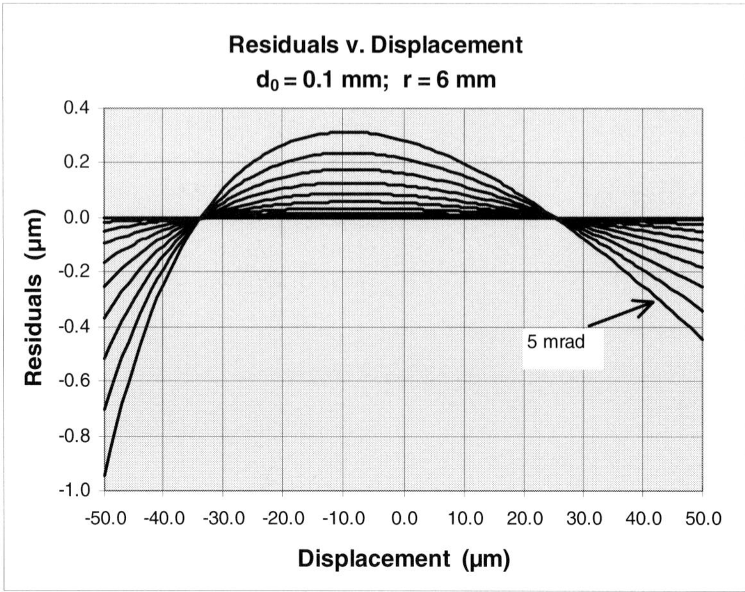

Fig. 5.10. Residuals at various tilt angles up to 5 mrad.
r=6 mm; d_0=100 μm; C=10 pF

More importantly, a constant tilt will effect the linearity of the measurement, Fig. 5.10 shows a family of residual curves for various tilt angles. In this plot tilt varies from 0 to 5 mrad. What does this mean for the user? A sensor as supplied will have been calibrated at zero tilt to have a scale factor of unity and linearity within the quoted specification for that sensor, i.e. the measured displacement is the same as the true displacement (within the calibration errors: see Appendix to Chapter 7). If then the sensor is mounted with a tilt angle, it will no longer be linear: Fig. 5.10 shows the residuals one could expect and

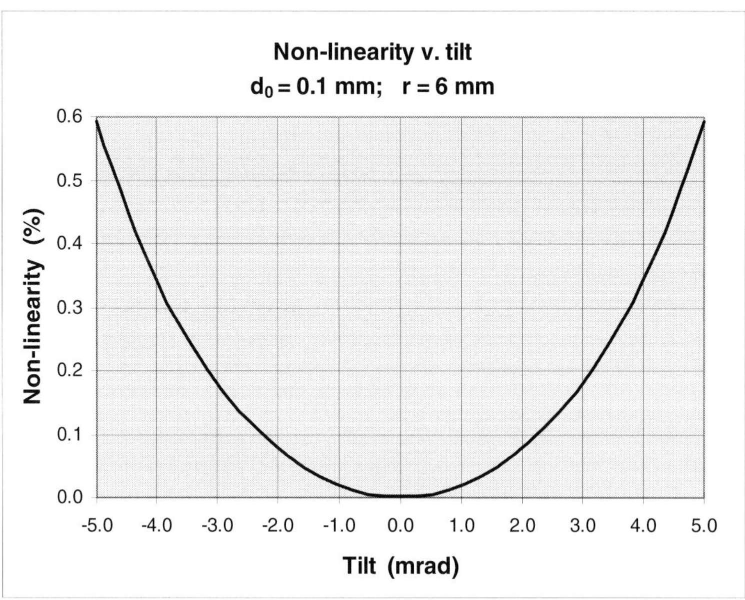

Fig. 5.11. Non-linearity v. tilt
r=6 mm; d_0=100 μm; C=10 pF

Fig. 5.11 shows the resultant variation of linearity with tilt. The non-linearity can be quite large: 5 mrad tilt gives about 0.6% non-linearity (remember, non-linearity is calculated as half the peak to peak value of the residual curve expressed as a percentage of the range).

Also the scale factor will differ from unity. Fig. 5.12 shows how the scale factor changes with tilt, as tilt is increased the scale factor increases. This means that a true displacement of, say, 1 μm will be measured as something larger than 1 μm if there is a tilt. Introducing a tilt also affects the nominal zero point of the measurement. The other curve of Fig. 5.12 shows this.

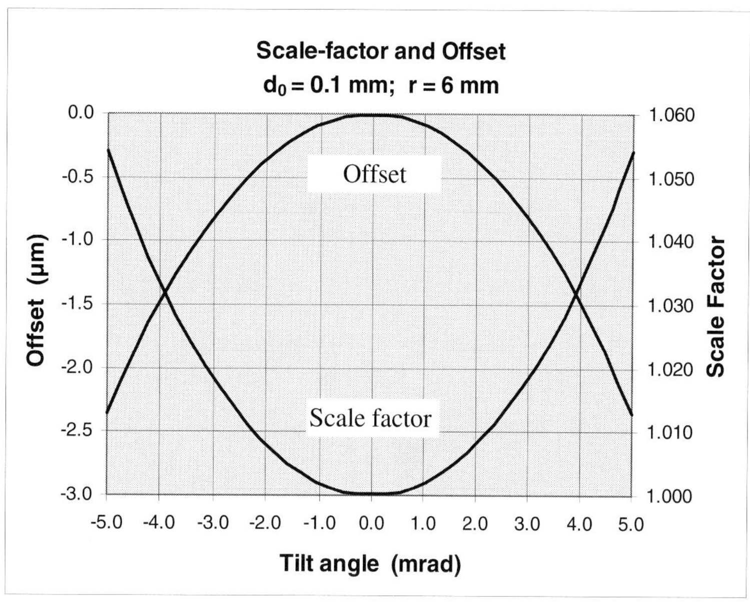

Fig. 5.12. Scale-factor and offset v. tilt r=6 mm; d_0=100 μm; C=10 pF

These plots are for our test capacitor of 6 mm radius and 100 μm nominal gap. In general, sensor capacitors with larger nominal gaps will be affected less by tilt as the gap change from centre to edge is less when expressed as a fraction of the nominal gap. This is shown in Fig. 5.13.

Fig. 5.13 shows the variation in non-linearity with nominal gap for a measurement span of 100 μm and constant tilt angle of 5 mrad. Two plots are shown: constant area and constant capacitance. The constant area plot is for our standard test capacitor with 6 mm radius, the nominal capacitance will vary from 10 pF to 1 pF as the nominal gap varies from 100 μm to 1 mm. The constant capacitance plot shows the variation that would be obtained if the radius of the electrodes were adjusted at each nominal gap to give 10 pF capacitance. The radius would have to vary from 6 mm at 100 μm nominal gap to 19 mm at 1 mm nominal gap.

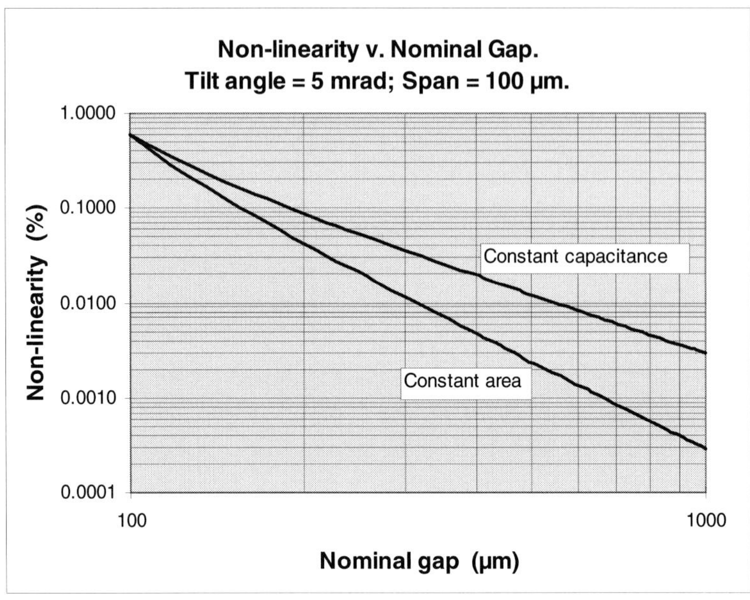

Fig. 5.13. Non-linearity v. d_0 at constant capacitance (10 pF) and constant electrode radius (6 mm).

As can be seen, the non-linearity for a given measurement span and tilt reduces enormously if the nominal gap is made larger, especially if the electrode radius is kept constant and the nominal capacitance is allowed to fall. Of course there would be a price to pay for doing this: noise. More of this later.

Plate bend

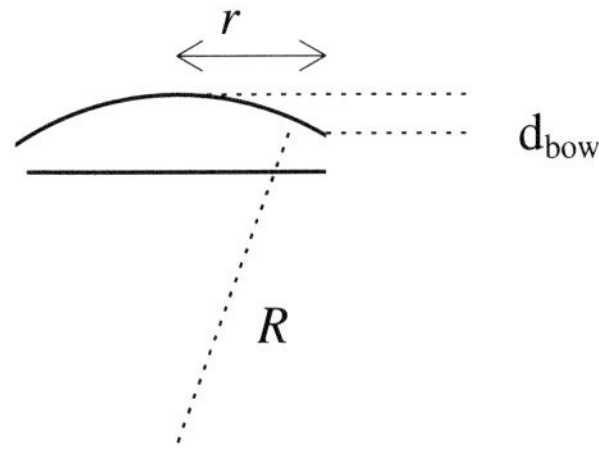

Fig. 5.14. Bowed capacitor

Just as tilt of one plate with respect to the other causes capacitance change and non-linearity, so will a bending of one plate with respect to the other. Fig. 5.14 illustrates a capacitor with one plate spherically bowed. The amount of bending is measured by the difference between the centre and edge gaps. A negative 'bow' indicates one plate is concave: the gap is larger in the centre as shown and R is considered negative. If the gap at the edge is greater than at the centre, then the bow is positive and R is considered positive.

The capacitance of this configuration can be found by integration of infinitesimal parallel plate capacitors as described in the Appendix. In this case

$$C = 2\pi\varepsilon_r\varepsilon_0 \int_0^r \frac{x}{d_0 \pm R \mp \sqrt{R^2 - x^2}}\mathrm{d}x \qquad (5.31)$$

and

$$d_{bow} = R - \left(\sqrt{R^2 - r^2}\right)\mathrm{sgn}(R) \qquad (5.32).$$

sgn(R) is -1 if R is negative; 0 if R is zero and +1 if R is positive. After suitable manipulation we have

$$C = 2\pi\varepsilon_r\varepsilon_0\left[d_{bow} - (d_0 + R)\ln\left(1 - \frac{d_{bow}}{d_0}\right)\right] \qquad (5.33).$$

Defining the plate curvature as $1/R$, Fig. 5.15 is a plot of capacitance v. bow given by the curvature varying from -0.2 m^{-1} to +0.2 m^{-1}. This gives a maximum bow of around 36 µm.

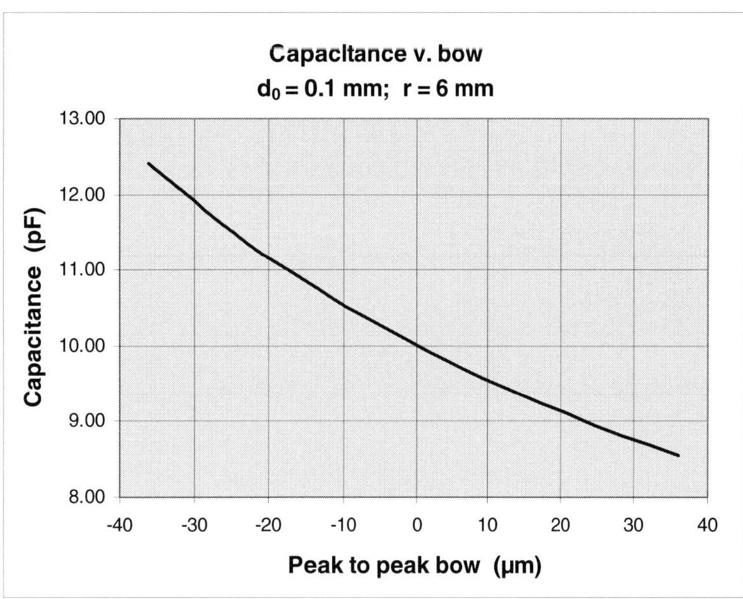

Fig. 5.15. Capacitance v. bow
r=6 mm; d₀=100 µm; C=10 pF

Figs. 5.16 and 5.17 show the variation of non-linearity with bow and variation of scale factor and offset with bow. The residual curve is similar to Fig. 5.10. Just as the case for tilt, linearity improves if the nominal gap is made larger.

Fig. 5.16 shows that the non-linearity can be quite large if the plates are seriously bowed, especially if they are concave.

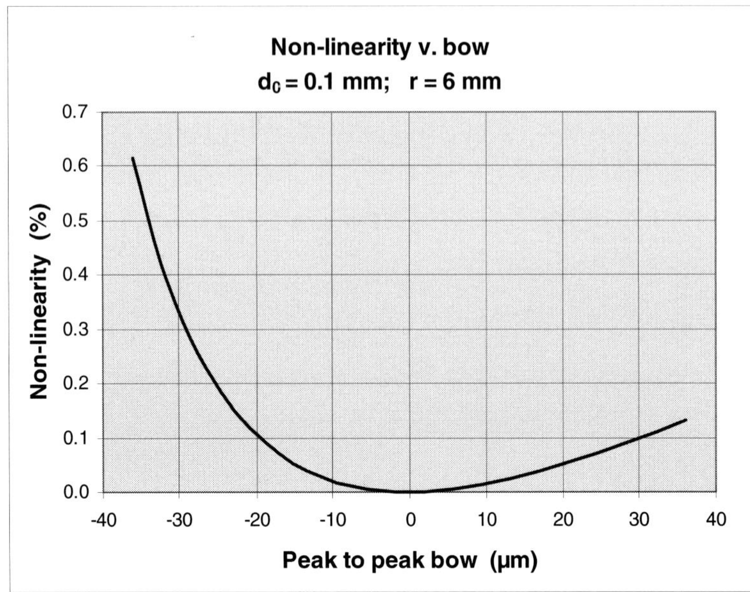

Fig. 5.16. Non-linearity v. bow
r=6 mm; d₀=100 µm; C=10 pF

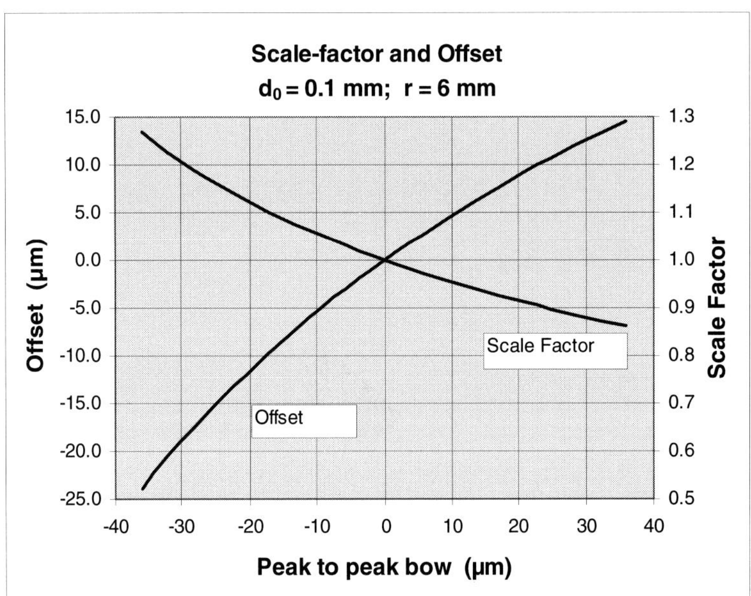

Fig. 5.17. Scale factor and offset
v. bow
r=6 mm; d₀=100 µm; C=10 pF

NOISE

If the plates of a capacitance sensor are held perfectly stationary at a fixed distance apart, the measured output will fluctuate due to electronic noise. This has been discussed in Chapter 2. As a rough rule of thumb, the rms measured noise density ($pm \cdot Hz^{-1/2}$) will be around 40×10^{-9} *of the nominal gap* for a 10 pF nominal capacitance. This scales with nominal capacitance, that is if the capacitance is doubled at a given nominal gap (by making the plates larger) the noise will be halved.

This does not bode well if one wishes to improve linearity by making the nominal gap larger. Doubling the gap from say 100 μm to 200 μm improves linearity for a tilted capacitor by over ten times but the nominal capacitance would halve. This doubles the noise coefficient and of course it is expressed as a fraction of the nominal gap which has also doubled. The noise is thus four times worse. For a fixed plate area the noise expressed as a displacement increases with the square of the nominal gap; for fixed capacitance it increases linearly with the nominal gap. Thus we have at fixed capacitance

$$d_{xp.\text{ndens}} = k_{xp.\text{ndens}} d_0 \Big|_{C_{\text{nom}}=10\text{pF}} \tag{5.34}$$

and at fixed plate area

$$d_{xp.\text{ndens}} = k_{xp.\text{ndens}} \frac{d_0^{\,2}}{d_{10pF}} \Big|_{r=6\text{mm}} \tag{5.35}$$

where

$k_{xp.\text{ndens}} =$ noise coefficient $= 40 \times 10^{-9}$
$d_{10pF} =$ gap for 10 pF $= 100$ μm.

These functions are plotted as Fig. 5.18. Note that this gives noise *density* as a function of gap, this must be multiplied by the square root of the bandwidth to get actual noise in nm.

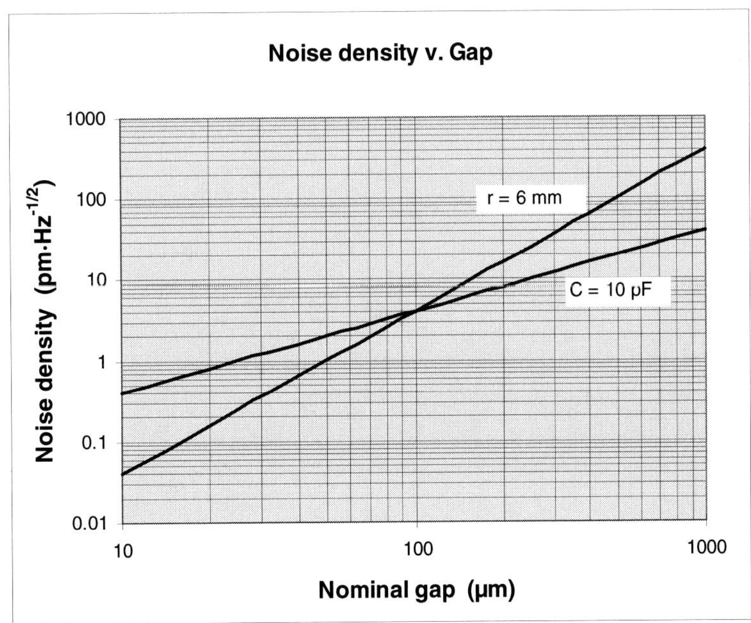

Fig. 5.18. Noise density v. gap
Constant r and constant C

It is worth noting that the noise coefficient is approximately that expected from electron shot noise at the current levels being measured. Capacitance micrometry could be improved by making the charge on the electron somewhat smaller!

USING SENSORS

So far in this section we have considered the nature of capacitance sensors and some of the problems that could be encountered when using them, concentrating on the 'short range' mode of operation where the displacement to be measured varies the capacitor gap. What other things must be taken into consideration?

Choosing a Sensor

Queensgate's standard 'NanoSensor' comes as a pair of plates denoted 'Target' and 'Probe' intended for use with the NPS range of sensor electronics and controllers. The probe has a circular active electrode of radius 6 mm and is surrounded with a guard ring at a gap of around 25 μm. The target has a larger radius than the probe and together the probe and target form the configuration shown in Fig. 5.7a. The plate surfaces are flat to a few micrometres.

The controllers can be set up to work with a range of nominal gaps, but 100 μm and 500 μm are standard. This gives nominal capacitances of 10 pF and 2 pF. As standard the measurement range is the same as the nominal gap going from $0.5d_0$ to $1.5d_0$, but again other ranges can be set up.

Range, linearity and noise

One should operate with the largest nominal gap consistent with the resolution required. The resolution will in general be dominated by the noise, so for example, if a noise density of around $0.1\ \mathrm{nm \cdot Hz^{-1/2}}$ is acceptable, one could work at 500 μm nominal gap giving a 500 μm measurement range with good immunity to tilt errors. The noise density of $0.1\ \mathrm{nm \cdot Hz^{-1/2}}$ would give a true noise displacement of around 3.2 nm with a set bandwidth of 1 kHz. Of course the bandwidth can be chosen to suit requirements: if the displacement to be measured only varies slowly one can set a low bandwidth, say a few hertz and get very high resolution (good precision). If on the other hand the displacement varies rapidly, a high bandwidth must be set and the resolution will suffer.

If it is necessary to measure with high precision over a short range, then work with the smallest gap consistent with the range and at the lowest bandwidth possible. The standard sensors will give around $0.2\ \mathrm{pm \cdot Hz^{-1/2}}$ with a nominal gap of 20 μm, so with a bandwidth of 10 Hz one could measure down to about 1% of a hydrogen atom radius! Of course to get good linearity one would have to ensure that the plates were mounted very parallel. Note that this example is entirely consistent with the NPS readout resolution of 1 part in 10^7.

Mounting considerations

When monitoring displacements over a long period of time, it is essential not to induce any stresses into the sensors when mounting them. These stresses will relieve over time and cause drift. Stresses can be minimised by using spring washers under screw heads (do not over tighten!) or in some situations by using magnets to hold the sensors in place. Queensgate will be happy to advise on particular mounting requirements.

The NPS range of controllers are designed to be immune to capacitance to ground and thus cable capacitance. It is, however, good practice to tie the cables down in such a way as to prevent excess cable motion and stress on the sensor heads.

Environmental effects

At the sort of resolutions that are within the capability of these sensors, vibration can be a dominant source of noise. In high precision applications it is essential to isolate the system being monitored from ground vibration and acoustic noise. Low frequency airborne noise from an air conditioning unit may not be audible but can seriously degrade performance.

During the course of development of the NPS range of controllers, we encountered an unexpected source of interference. After a lot of messing around, it was traced to the cooling fan we were using. The supplier seemed unconcerned, so we changed our supplier.

External electro-magnetic interference can also be a source of problems. Whereas Queensgate's equipment meets the stringent requirements of the European directives on electro-magnetic compatibility, this still gives a lot of scope for measurement errors in the presence of high radiated fields (one must remember, we are dealing with just a few electrons sloshing around!). Cables etc. are well shielded to minimise this problem but it is prudent to keep the sensors away from potential field sources such as computer monitors or noisy electric motors.

The sensors work by measuring the capacitance of the capacitor formed by the target and probe sensor plates. Unfortunately this capacitance depends not only on the plate area and gap but also on the relative permittivity , ε_r, of the air between the plates: a 1×10^{-6} change in the relative permittivity will cause an output equivalent to a 1×10^{-6} change in the nominal spacing.

The value of ε_r depends on the partial pressures of dry air, carbon dioxide and water vapour in the atmosphere. A useful equation is

$$(\varepsilon_r - 1)\times 10^6 = \frac{1553.9}{T}p_1 + \frac{2663.6}{T}p_2 + \frac{1295.2}{T}\left(1 + \frac{5748}{T}\right)p_3 \qquad (5.36)$$

where

T = absolute temperature (K) (300 K)

p_1 = partial pressure of CO_2 -free dry air (kPa) (101.3 kPa)

p_2 = partial pressure of CO_2 (kPa) (0.035 kPa))

p_3 = partial pressure of water vapour (kPa)
(3.567 kPa at 100 %RH and 300 K)

This is derived from the work of L.Essen and K.D.Froome (1951) on the refractive index of air.

These effects are in the order of a few $x10^{-6}$ for typical atmospheric variations. Humidity, however, has a much greater effect than would be expected. From equation 5.36 the variation expected is about $3x10^{-6}$ per one percent change in relative humidity (change of 0.036 7 kPa) or 1.5 nm·%$^{-1}$. In practice larger changes than this are observed due to adsorption of water vapour onto the plates. Please contact Queensgate if this is likely to be a problem.

Making your own sensor

Queensgate's standard NanoSensors are suitable for a wide variety of applications, but it is sometimes necessary to customise designs for a particular use. Often this is quite straightforward: after all a capacitance sensor is just two conducting plates! Queensgate will be happy to advise on sensor design or, of course, design a special. Key points to watch are:

1. use the maximum plate area possible;

2. make the plates out of a chemically inert conductor;

3. decide on the required noise density and use the maximum nominal gap consistent with achieving it;

4. use a guard ring on the probe electrode;

5. ensure the plates can be mounted without stressing themselves or the device they are to be mounted on;

6. ensure that the plates can be mounted parallel and that they are not bent;

7. shield the sensors from electro-magnetic fields;

8. isolate the system from mechanical and acoustic vibration.

General capacitance determination

Laplace's equation and other hard sums.

In principle the capacitance of any configuration of electrodes can be determined by solving Laplace's equation

$$\nabla^2 V = 0 \qquad (5.37)$$

for the configuration involved. V is the potential at any point in the region of the conductors. Once V as a function of position is determined, the surface charge density on an electrode is given by

$$\sigma = -\varepsilon_r \varepsilon_0 \left(\frac{\partial V}{\partial z} \right)_{z=0} \qquad (5.38)$$

where z is the dimension perpendicular to the electrode surface. Once the surface charge density is known, it can be integrated over the surface to give the charge

$$Q = \int_A \sigma \, \mathrm{d}A \qquad (5.39).$$

Knowing the charge and the voltage difference between the electrode in question and the other electrode to which the capacitance is to be determined, the capacitance is then simply given by equation 5.1. If all this sounds rather difficult, that is because it is: in fact it is impossible to solve analytically for most configurations one is interested in!

Things get a bit easier for circular plate configurations where the potential V does not depend on the azimuth angle ϕ. The Laplace equation can then be written in cylindrical polar coordinates

$$\frac{\partial^2 V}{\partial r^2} + \frac{1}{r} \frac{\partial V}{\partial r} + \frac{\partial^2 V}{\partial z^2} = 0 \qquad (5.40)$$

which can be solved by using separation of variables and conformal transformations and applying the boundary conditions dependent on the capacitor geometry. Heerens and Vermeulen (1975) have done some splendid work on this, applied to guard ring configurations.

A more practical approach these days is to let the computer do the work. Equations 5.37 to 5.39 can be solved for any geometry using finite element analysis techniques. This approach is adopted by Queensgate for the analysis of difficult configurations.

Alternatively one can get a good idea of the capacitance of most arrangements by integrating the capacitance of infinitesimally small 'perfect' capacitors over the area of the electrodes, i.e.

$$C = \varepsilon_r \varepsilon_0 \int_A \frac{\mathrm{d}A}{d} \tag{5.41}.$$

This cannot be used to assess edge effects but is useful for analysing gross defects like tilt and bending.

Expressions for the effect of tilt

Expressions have been derived for the C term of equation 5.13 with respect to tilt angle θ by Heerens and Vermeulen (1975) using solutions to Laplace's equation; Harb *et al.* (1995) by analytical integration of equation 5.41 and Hicks T.R. (1995) by numerical integration of equation 5.41. The three expressions as presented are different but all give the same answers to a few parts per million. Zhao, Wilkening, Marth and Horn (1994) quote a formula that appears to be an integration of equation 5.41 but it is wrong, giving imaginary answers. The error seems to be typographical: a cosine where there should be a sine and a missing factor of two. Making these 'corrections' their result becomes identical to that of Harb *et al.* (1995).

<u>Heerens and Vermeulen</u>

$$C = \frac{\varepsilon_r \varepsilon_0 \pi r_i^2}{d} \left[1 + \left(\frac{1 - \sqrt{1 - \left(\frac{r_i \theta}{d} \right)^2}}{1 + \sqrt{1 - \left(\frac{r_i \theta}{d} \right)^2}} \right) \right] \tag{5.42}.$$

<u>Harb *et al.*</u>

$$C = \frac{2\varepsilon_r \varepsilon_0 \pi r_i^2}{d} \left(\frac{1}{1 + \sqrt{1 - k^2}} \right) \tag{5.43}$$

where

$$k = \frac{r_i \sin 2\theta}{2d} \tag{5.44}.$$

<u>Hicks T.R.</u>

$$C = 2\varepsilon_r \varepsilon_0 \int_{-r_i}^{r_i} \frac{\sqrt{r_i^2 - x^2}}{d + x \sin \theta} \mathrm{d}x \tag{5.45}.$$

It will be noted that if θ is small then equations 5.42 and 5.43 are identical.

REFERENCES

Brown M.A. and Bullied C.E. The effect of tilt and surface damage on practical capacitance displacement transducers *Journal of Physics E: Scientific Instruments Vol.11, 1978*

Essen L. and Froome K.D. The refractive indices and dielectric constants of air and its principle constituents at 24,000 Mc/s. *Proceedings of the Physical Society, LXIV, 10-B 1951*

Heerens W.Chr. and *Vermeulen F.C.* Capacitance of Kelvin guard-ring capacitors with modified edge geometry. *Journal of Applied Physics, Vol 46, No.6, June 1975.*

Heerens W.Chr. Applications of capacitance techniques in sensor design, *Journal of Physics E: Scientific Instruments Vol.19, 1986.*

Hicks I. Private communication. *1997*

Hicks T.R. Queensgate Instruments Report *1995*

Jones R.V. and *Richards J.C.S.* The design and some applications of sensitive capacitance micrometers. *Journal of Physics E: Scientific Instruments Vol.6, 1973.*

Maxwell J.C. A treatise on electricity and magnetism, *Oxford Clarendon 1873*

Moon C. and Sparks C.M. Journal of research of the National Bureau of Standards 41, 1948

Scott A.H. and Curtis H.L. Journal of research of the National Bureau of Standards 22, 1939

Harb S.M., Chetwynd D.G., Smith S.T. Tilt errors in parallel plate capacitive micrometry. *Bonis M., Alayli Y., Revel P., McKeown P.A.,Corbett J., (eds) International Progress in Precision Engineering (Elsevier), 147-50: 8th International Precision Engineering Seminar, Compiègne, May 1995*

Zhao X., Wilkening G., Marth H. and *Horn K.* High accuracy capacitance sensor for positioning of piezoelectric displacement actuator. *Actuator 94, Bremen, Germany, June 1994*

It has been known for a long time that certain naturally occurring minerals, for example quartz, will produce a voltage if they are subjected to stress. It has also been known that the same minerals will change dimension if an electric field is applied to them. This class of behaviour is known as the piezoelectric effect and is very useful for producing small displacements.

In this Chapter:

SOME PROPERTIES OF PIEZO MATERIALS

The piezoelectric effect is small in naturally occurring minerals, but present day ceramic technology has given us a range of man-made materials that do a lot better. Most of these are based on lead zirconate titanate and are generally known as PZT ceramics (from Pb, Zr, Ti). These can produce strains of up to 0.1% enabling motions in the order of 100 μm with a device 100 mm long. Strictly speaking, these are ferroelectric materials and have to be polarised before they function. This is done by subjecting the material to a high field, usually at high temperature.

Actuators made using piezoelectric materials (piezo actuators for short) have a number of advantages over other forms of actuator:

- the motion is smooth and continuous - no stick-slip (the expansion is an atomic process). In practice the smallest step size is limited by the noise level of the controller.

- Piezo actuators are very stiff: practical actuators have over 20 % of the stiffness as the equivalent made out of solid stainless steel. They can thus generate a lot of force.

- Piezoelectric ceramic responds quickly to step inputs. In practice the response is limited by the current rating of the controller.

- They have low power dissipation, especially when static. Typical power dissipation: moving, a few milliwatts; static, a few microwatts.

Piezo actuators have their problems too:

- they are notorious for being nonlinear and suffering large hysteresis and long term drift.

- High strain materials have a low Curie temperature: the Curie temperature is the temperature at which polarisation is lost and the actuator ceases to function.

- They are not very good at pulling (though they will do it) so require pre-loading.

Of course the non-linear and hysteretic behaviour can be overcome if the actuator is used in a closed loop system with a NanoSensor providing feedback, but it is instructive to look at the open loop properties.

Large field properties

These are best discussed with reference to the 'butterfly' diagram of Fig. 6.1 which shows a plot of extension v. voltage over a large range.

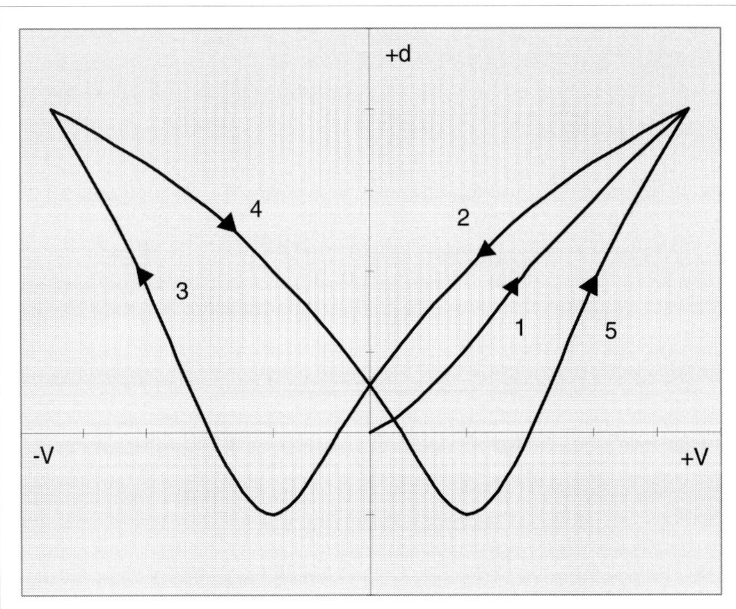

Fig. 6.1. The butterfly diagram

Starting with un-poled material at the origin of the diagram, as the voltage is increased, the material will start to pole and expand, following path 1. It will eventually reach a saturation extension at high positive voltage. If then the voltage is decreased, the extension will follow path 2 down to some negative voltage where the original polarisation will be lost and the material will re-polarise in the opposite direction. It will then expand again as the voltage becomes more negative, following path 3, again to a saturation point. Reversing the voltage again makes the extension follow path 4 back to a positive voltage that polarises the material in the original direction again. It then follows path 5 back up to the original saturation point from where it would follow path 2 again. This can hardly be called linear behaviour!

Low field properties

In practice piezo actuators are used over a smaller voltage range than required to produce Fig. 6.1 (especially in the negative direction) and one starts with poled material. The extension v. voltage curve is more like Fig. 6.2.

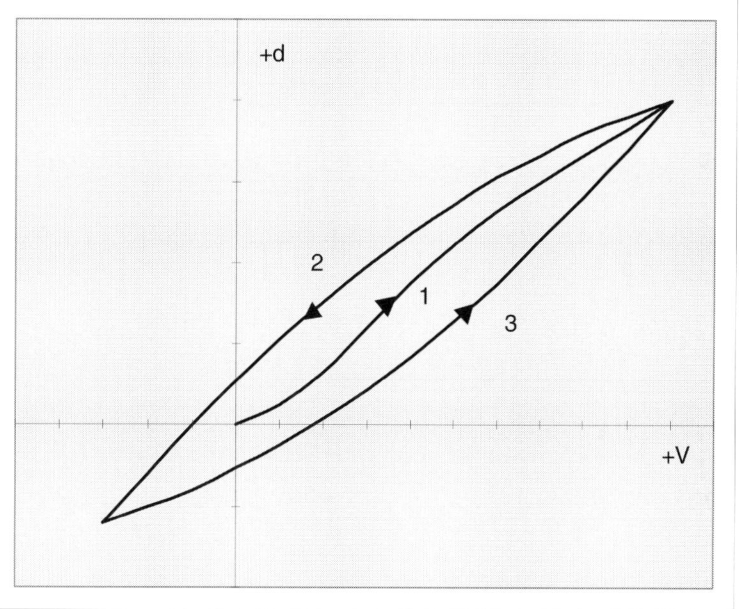

Fig. 6.2. Normal operating voltage

Starting now with a poled actuator with zero volts applied, the rather indeterminate path 1 will be followed as the voltage is increased. When the voltage is reversed, path 2 will be followed. Assuming that the voltage is not taken negative enough to de-pole the material, path 3 will be followed when the voltage is taken more positive again. If the voltage is cycled between two fixed levels, the extension will follow the closed loop defined by paths 2 and 3.

Hysteresis

As can be seen from Fig. 6.2, there is a large hysteresis. This is defined as the maximum difference between the upward and downward paths when the actuator is cycled repeatedly over its full range, expressed as a percentage of the full range. Values of 20% are not uncommon. See

Chapter 2, Accuracy, Trueness and Precision, *Position Reproducibility* for further discussion of hysteresis.

Linearity

Linearity can not be defined for a piezo in the same way as it can for, say, a sensor as the large hysteresis dominates: the peak to peak deviation from the straight line fit would measure hysteresis and not non-linearity. A somewhat different approach is required.

Fig. 6.3 shows another hysteresis loop. If the actuator travels the large loop represented by points A and E, it can be said to have a scale factor defined by the line AE. If now a smaller loop were traversed, represented by points A and D, the scale factor would be less, represented by the line AD. The return paths for this and other smaller loops A-C, A-B are not shown for clarity. As can be seen, as the loops get smaller, so the scale factor reduces.

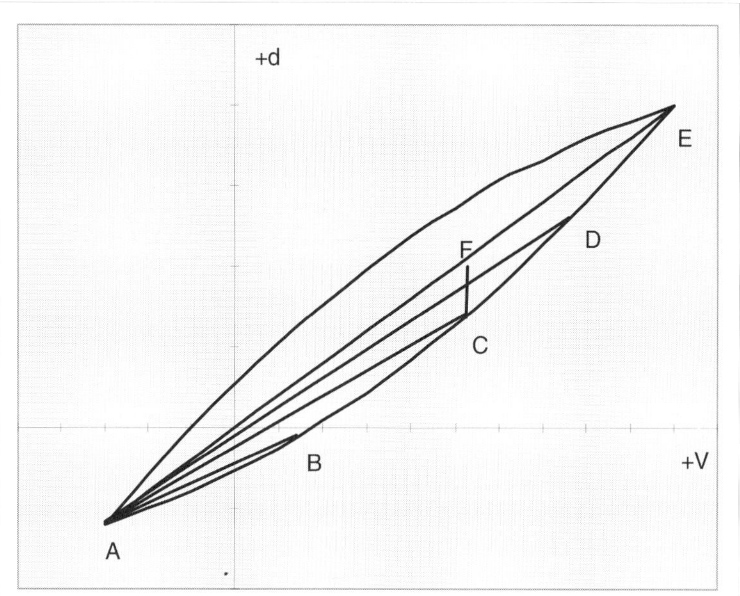

Fig.6.3. Piezo non-linearity

This effect does not just happen when one starts from the same point A: smaller loop amplitudes give lower scale factors wherever one starts from. The piezo non-linearity thus manifests itself as a variation of the scale factor over the range for cyclic applied voltages. The small signal scale factor can be one third to one half of the large signal.

Creep

Looking again at Fig. 6.3, if one were to follow the cycle defined by points A and C and then stop at point C, the extension would 'creep' up to point F somewhere close to the line for the maximum loop. This can take several tens of minutes. It could thus be said that the real dc scale factor, allowing the piezo to creep to its final extension, is approximately constant over the range. Once at F, if the cycle A-C was resumed using the same limit voltages as originally, the first half-cycle (or maybe one and a half cycles) would take the extension from F to A and then the A-C loop would be followed again. Piezos are weird devices!

PRACTICAL ACTUATORS

Construction

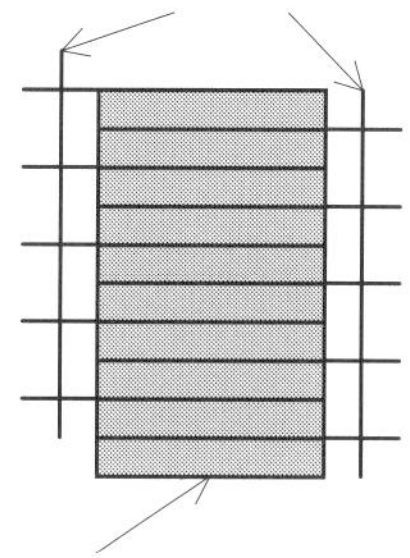

Fig. 6.4. A section of a piezoelectric actuator

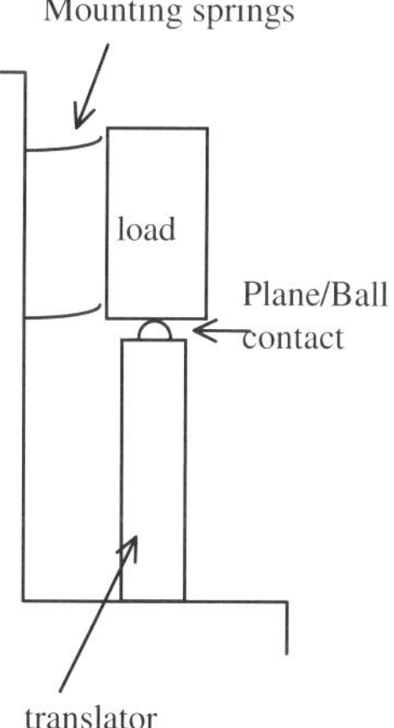

Fig. 6.5. Ball contact

Fig. 6.5 shows the actuator mounted vertically but it does not have to be. This is just convenient to illustrate the effect of a gravitational load and also the diagram fits in the margin better.

Quite high electric fields are required to get the materials to expand, so piezo actuators are made up of thin ceramic elements interleaved with conducting electrodes: see Fig. 6.4. The ceramic elements are mechanically in series but electrically in parallel, this enables full extension to be obtained with as little as 150 V or even 60 V with some devices.

Pre-load

Piezo materials are very good at pushing but not too good at pulling. Also they are very strong in compression but rather weak in tension. They will contract when a negative voltage is applied (as shown in the preceding diagrams) and can indeed generate some force while doing so, but there is a danger of breakage. For this reason pre-loading is a good idea.

Pre-loading consists of applying a constant compressive force to the stack of elements using a spring assembly. This keeps the ceramic elements under compression even when the actuator is pulling on a load, as long as the required pull force is less than the pre-load. Pre-loads of a few hundred newtons are not uncommon.

Queensgate's range of piezo actuators are constructed from stacks of PZT ceramic elements housed in a cylindrical steel shell containing the pre-load springs. Drilled and tapped steel end pieces on the stack enable easy mechanical interfacing.

Mounting

Piezo materials contract in the radial direction as they expand in the axial direction. Also the axial expansion may not be uniform over the area of the elements due to material composition and poling inhomogeneity. These effects conspire to make the ends of the stack tilt and twist in unexpected directions with respect to each other as the actuator expands, in fact on the nanometre scale they tend to resemble piezoelectric corkscrews. These nasty effects can be decoupled from the actuator end pieces using flexure mechanisms (see Chapter 7 NanoMechanisms) but are present in open-loop actuators. The best way to mount a actuator is thus to fix it at one end and get it to apply force to the load via a ball on the other, see Fig. 6.5. In this way lateral or angular motions of the end are decoupled from the load.

If the actuator *must* be fixed at both ends, then take care not to exceed angular or radial load specifications, especially while installing the actuator.

Static performance

Piezo actuators can develop a large amount of push force (100's or 1000's of newtons) but they are not infinitely stiff. This can affect the amount of motion one actually ends up with.

Constant load

Take an example of a actuator capable of moving 15 μm with no load and having a stiffness of 50 N·μm^{-1}. If a weight of 10 kg is applied, say using the configuration of Fig. 6.5 with weak mounting springs, the actuator will be squeezed by about 2 μm: 10 kg weight is about 100 N. If now the voltage for full range is applied to the actuator, it will still expand by 15 μm, but from its shifted start point. A constant load will not affect the range. In this example it is assumed that the mounting spring force does not vary over the 15 μm range. Of course one must not exceed the maximum load of the device: too much compressive force will de-pole the material (before it crumbles).

Spring load

Consider now the case of Fig. 6.5 but with strong mounting springs. As the actuator expands, the force on it due to the springs will increase. This increase in force reacts with the stiffness of the actuator to push it back a bit so the total range for a given applied voltage is reduced. If the actuator has a stiffness of $k_{x.\text{trans}}$ and the springs have a total stiffness of $k_{x.\text{spring}}$, then the resultant extension d_x is given by

$$d_x = \frac{k_{x.\text{trans}}}{k_{x.\text{trans}} + k_{x.\text{spring}}} d_{x.\text{max}} \tag{6.1}$$

where $d_{x.\text{max}}$ is the no-load extension of the actuator. Obviously some potential range will be lost because of the actuators internal pre-load springs, but this is taken into account when the maximum range of the actuator is specified.

Dynamic Performance

Resonance

The stiffness of the actuator also affects the dynamic performance when trying to move a mass. The stiffness and the mass combine to form a resonant system with natural frequency

$$f_0 = \frac{1}{2\pi}\sqrt{\frac{k_{x.\mathrm{trans}}}{m_{\mathrm{load}}}} \tag{6.2}.$$

Our example above with a 10 kg load would thus resonate at about 350 Hz, so it would not respond well to frequencies higher than this: a stiffer actuator would be required to move faster. The no-load resonant frequency of the actuator can be tens of kilohertz. In practice though it is often the drive amplifier characteristics that limit the speed, read on.

Slew Rate

As will have been noticed, the construction of a piezo actuator is very similar to a ceramic capacitor, in fact some actuators are manufactured using ceramic capacitor fabrication techniques. They *are* capacitors and quite large ones too: the ceramic properties that enable high strains to be developed also give the material a high dielectric constant. Capacitor values of a half to a few microfarads can be expected. The amplifier driving the piezo will have a limited current output capability and this limits the rate of change of voltage that can be developed across the piezo. If the amplifier drive current is I_{drive}, then

$$\frac{\mathrm{d}V_{\mathrm{drive}}}{\mathrm{d}t} = \frac{I_{\mathrm{drive}}}{C_{\mathrm{pzt}}} \tag{6.3}$$

where C_{pzt} is the actuator capacitance. If S_{pzt} is the piezo voltage coefficient ($\mu\mathrm{m}\cdot\mathrm{V}^{-1}$) then the maximum extension rate is

$$u_{x.\max} = S_{\mathrm{pzt}}\frac{I_{\mathrm{drive.max}}}{C_{\mathrm{pzt}}} \tag{6.4}.$$

Velocities of 5 $\mu\mathrm{m}\cdot\mathrm{ms}^{-1}$ are typical. The limits this places on the maximum sine wave amplitude that can be handled have been discussed in Chapter 3, Servo Control, *Response Curves and Settling Time*

Power handling

Perfect capacitors pass alternating current without any power dissipation: the current is always 90° out of phase with the voltage so the *IV* product integrates to zero in a cycle. However in the output stages of the drive amplifier, the current and voltage are definitely not

$90°$ out of phase and there is power dissipation there. Returning to the piezos: piezos are capacitors but they are not perfect capacitors. If the piezo is driven with a sine wave of peak to peak amplitude V_{drive} at a frequency f, there will be a dissipation given by

$$P_{pzt} = \frac{\pi . f . V_{drive}^2 . C_{pzt} . \tan\delta}{4} \tag{6.5}$$

where $\tan(\delta)$ is the dissipation factor for the material and C_{pzt} is the capacitance of the piezo.

A typical value for $\tan(\delta)$ would be 0.02, so our example of a $1\,\mu F$ actuator with a 10 kg load and peak to peak drive of 150 V would dissipate about a quarter of a watt at the resonant frequency of 360 Hz. This would be within the safe operating power for the actuator, but if the 10 kg load were removed, the possible operating frequency could go up to several kilohertz. Dissipation could then be a few watts which would cause over-heating and de-poling of the piezo material.

Power dissipation in the drive amplifier can be much higher than in the actuator

$$P_{amp} = C_{pzt} . V_{drive} . V_{max} . f \tag{6.6}$$

where V_{max} is the maximum output capability of the amplifier (peak to peak). Our example at 360 Hz would give a drive amplifier dissipation of around 8 W which must be allowed for in drive amplifier construction. As you can see, one needs a rather beefy amplifier to drive actuators fast!

As well as power dissipation in the amplifier one must consider its maximum current drive capability. The peak output current depends on the drive waveform

$$I_{drive.avg} = C_{pzt} . V_{drive} . f \tag{6.7}$$

$$I_{drive.pk.sine} = \pi . C_{pzt} . V_{drive} . f \tag{6.8}$$

$$I_{drive.pk.triangle} = 2 . C_{pzt} . V_{drive} . f \tag{6.9}.$$

The average current is independent of waveform.

Noise and Resolution

The resolution of piezo actuators is governed by the noise on the drive amplifier rather than any property of the piezo: there is no intrinsic dead band in the material. The rms drive amplifier noise will be generally less than $3\,\mu V \cdot Hz^{-1/2}$ giving a noise displacement of $0.4\,pm \cdot Hz^{-1/2}$. This is usually adequate for most applications.

CHAPTER 7. NANOMETRE PRECISION MECHANISMS

Theory meets practice

In this chapter we bring together the concepts of the previous chapters in order to describe real nanometre precision mechanisms. We discuss how they work, how they are specified, how they are tested and give some hints for getting the best results from them.

> *...Theory is when you understand everything but nothing works.*
>
> *...Practice is when you don't understand anything but everything works.*
>
> *...When theory meets practice you don't understand anything and nothing works.*
>
> *Of course, Queensgate uses the obvious alternative last line!*

DESIGN PHILOSOPHY

Piezo-electric devices have the potential to move stages with the resolution required for nanometre precision mechanisms. However, as we have seen in the last chapter, an external sensor is required to control their position because piezo-electric devices are non-linear and exhibit hysteresis. The capacitance micrometer is ideally suited to this task, being small and simple with an intrinsic resolution capability which is effectively infinite. The motion measured by the sensor is fed back to the controller which moves the stage to minimise the difference between the sensed motion and the command. This form of controller has been described extensively in Chapter 3, Servo Control. As was shown, the positioning precision in the metrology loop is mainly determined by the capabilities of sensor and controller.

Guided motion

As has been stated in Chapter 6, Piezos, piezo actuators behave more like nano-corkscrews. To achieve a pure single axis motion, flexure guiding mechanisms are used to constrain off-axis motions and combine piezo actuator and sensor together to form an integral stage system. Fig. 7.1 is a typical closed loop control block diagram of this kind of system. Flexures may be parallel spring strip designs for linear motion or cross springs for angular motion (see Smith and. Chetwynd, 1992 for a discussion of these and their variants).

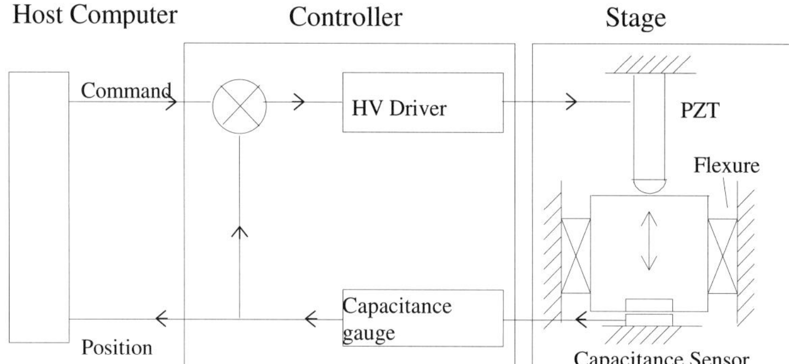

Fig. 7.1 Single axis stage control loop

In principle flexure mechanisms can provide very pure motion, but to do this they must be carefully designed and manufactured. The aim is to get low stiffness in the direction of the required motion and high stiffness in the other directions without introducing undue stresses and friction. Finite Element Analysis (FEA) is able to predict local and global distortions enabling the design to be optimised by decoupling the forces or making the inevitable distortions cancel each other. Friction must be avoided as it will introduce hysteresis, especially if it appears between the sensor and the point on the stage where the load is mounted. Inevitably there will be some parasitic motion, but careful manufacture ensures that this is repeatable and characterisable and therefore correctable: see *Using NanoMechanisms*.

Mechanical amplification

Piezos are great for producing small motions but this can also be a problem as their range is limited: it requires an actuator 100 mm long to achieve a range of 100 μm. This can be overcome by using a shorter actuator and a mechanical lever arrangement to amplify the motion. Here the piezo's high pushing force and stiffness properties are being used to provide a longer extension from a more compact device. The resultant stiffness of the piezo actuator will be reduced to

$$k_{\text{eff}} = \frac{k_{\text{pzt}}}{G_{\text{mech}}^2} \tag{7.1}$$

where k_{pzt} is the stiffness of piezo stack itself and G_{mech} is the mechanical amplification. Amplification factors of five or so are reasonable while still retaining useful stiffness. Higher amplification factors are possible but the stiffness, and thus the resonant frequency, becomes very low. Often the lever arrangement is fabricated as part of the guiding flexure mechanism.

Metrology frame and kinematic mounting

Motion must be measured relative to something and in a nanometre precision mechanism it is the mechanical frame assembly holding the 'stationary' plate of the capacitance sensors. If this moves, errors will be introduced into the motion. Kinematic or isostatic mounting of this metrology frame onto the user's system minimises induced stresses to and from the user's system.

Kinematic design is discussed extensively by Smith and Chetwynd (1992). Basically kinematics is the study of the geometry (direction) of motion, to be contrasted with kinetics, which studies the forces involved. Kinematic locations use one (and only one) point contact to constrain each degree of freedom, thus six (and only six) point contacts are required to locate a stage (x, y, z, γ, θ, ϕ). A common way of doing this is to use three balls rigidly attached to the metrology frame resting in three vee grooves in the user's mounting base, see Fig. 7.2a. Each ball has two contact points with the grooves, giving the required six in all. Gravity keeps the balls and grooves in contact. It can be seen from the arrangement that if the metrology frame were to expand with respect to the user's system, the balls could slide in the grooves and no forces would be transmitted.

The above kinematic mounting has a few problems:

1. it requires the user to machine grooves in the system baseplate. This may not be convenient.

2. It utilises gravity to hold things together. This could be replaced by a spring, but that entails yet another fixing to the user's system and would make installation difficult.

3. It relies on sliding contacts to decouple thermal expansion effects. These are difficult to predict and friction would induce stress in the metrology frame and the baseplate.

These problems can be solved by using spring strip arrangements in place of the ball in vee groove: see Fig. 7.2b. The spring strips are fashioned into the metrology frame of the stage using a pattern of thin cuts. This scheme allows mounting to the user's system baseplate with three screws through the holes provided. This is about as convenient as

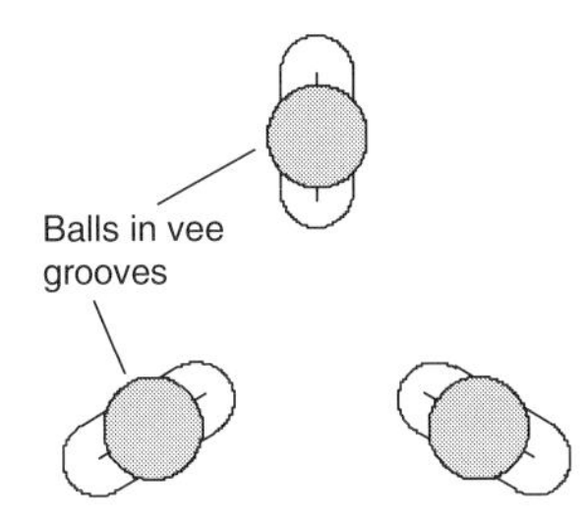

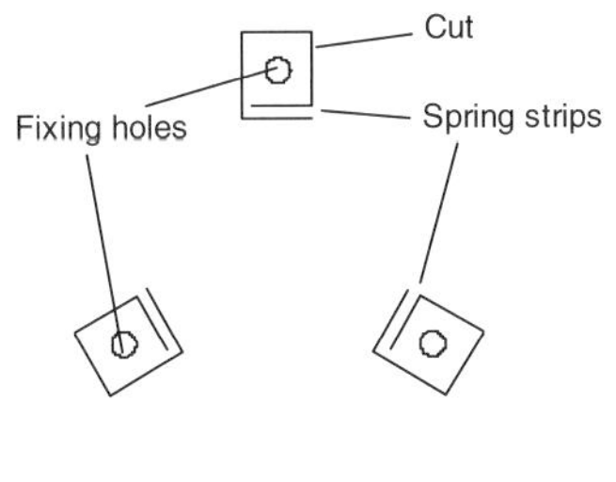

Fig. 7.2 a) Kinematic location.
b) Isostatic implementation

one can get for a precision mechanism and does not rely on gravity for constraint. It will be noticed that this mounting is not strictly kinematic as the stage is somewhat over constrained: a kinematic mounting point constrains motion in one direction while allowing it in all others, a spring strip allows motion in one direction while constraining it in others. Stresses due to thermal expansion are, however, properly decoupled if the allowed motion axes of each strip meet at the same point. This point then becomes the origin from which motions due to thermal expansion can be calculated and is conveniently arranged to be at the centre of the moving part of the stage.

This form of mounting is generally known as *isostatic,* which strictly means 'floating'.

SPECIFICATIONS AND PERFORMANCE MEASUREMENT

The truth is out there …

The perfect NanoMechanism has pure axial motion, high structural and thermal stability, high resolution, high resonant frequency, perfect step response, absolutely linear output and orthogonal motions for multi-axis systems. So what about the real world?

The performance of a stage can be summed up by its accuracy. So we just need one number to quantify accuracy, then throw in a few other details like settling time, weight and size and we have a specification sheet. Simple. Simple, yes, but not very helpful. It is far better to break down the accuracy into some of its component parts and specify these to enable the user to determine in detail whether the stage is suitable for the application. For example it may be very important for one application that the linearity be excellent but noise may not matter; another may require low noise, and so on. Also the user may be able to correct for some errors such as non-linearity and parasitic rotations if they are quantified. For these reasons the components that make up accuracy are specified, rather than the overall figure. Table 7.1 on page 85 shows a sample specification sheet for a X-Y stage.

We will now detail each of the specification parameters in turn. An entry in the 'Typical' column, or a value described as typical in the comments column, indicates a value commonly achieved but not guaranteed. Entries in the 'Maximum' or 'Minimum' columns are guaranteed.

Where applicable, parameters are measured for each axis of the stage. The procedures we use to measure these parameters are also outlined to help understand what they mean.

TYPICAL 100μm XY STAGE							
Parameter		**Symbol**	**Value**			**Units**	**Comments**
STATIC							
			Minimum	**Typical**	**Maximum**		
Material			Super Invar (Bright nickel plated)				
Size				100 x 100 x 23		mm	40 mm dia.Aperture
Material thermal expansion coefficient		α_{mech}		0.3		ppm·K^{-1}	
Range	X	$d_{x.\mathrm{max}}$	±50			μm	Note 2
	Y	$d_{y.\mathrm{max}}$	±50			μm	
Scale factor	X	b_{x1}		1		μm	Command units are micrometres
	Y	b_{y1}		1			
Scale factor error, position	X	δb_{x1}			0.1	%	
	Y	δb_{y1}			0.1	%	
Static stiffness	X	k_x	50			N·μm^{-1}	Note 1
	Y	k_y					
Allowable load		m_{load}	0		0.5	kg	Note 1
DYNAMIC							
			Fast	**Medium**	**Slow**		
Effective bandwidth	X	Bpx	800	100	10	Hz	3dB. Typical.
	Y	Bpy		100	10	Hz	Note 2
Lowest resonant frequency 0 g load		$f_{0.0\mathrm{g}}$		1000		Hz	Note 2
100 g load		$f_{0.100\mathrm{g}}$		500		Hz	
500 g load		$f_{0.500\mathrm{g}}$		100		Hz	
Small signal settling time	X	$t_{xs.s}$	0.2	1.6	16	ms	To ±2%. Typical.
	Y	$t_{ys.s}$	0.2	1.6	16	ms	Note 2
Large signal settling time	X	$t_{xs.l}$	16	18	32	ms	To ±2%. Typical.
	Y	$t_{ys.l}$	16	18	32	ms	Note 2
Slew rate	X	$u_{xp.\mathrm{max}}$	5	5	5	μm.ms^{-1}	Minimum.
	Y	$u_{yp.\mathrm{max}}$	5	5	5	μm.ms^{-1}	Note 2
Position noise	X	$\delta x_{p.n}$	0.11	0.04	0.013	nm	rms Maximum.
	Y	$\delta y_{p.n}$	0.11	0.04	0.013	nm	Note 2
MECHANICAL ERROR TERMS							
			Minimum	**Typical**	**Maximum**		
Hysteresis	X	$\delta x_{p.\mathrm{hyst}}$		0.0005	0.001	%	peak to peak.
	Y	$\delta y_{p.\mathrm{hyst}}$		0.0005	0.001	%	Note 2
Non-linearity	X	$\delta x_{p.\mathrm{lin}}$		0.02	0.05	%	peak.
	Y	$\delta y_{p.\mathrm{lin}}$		0.02	0.05	%	Note 2
Yaw	X	$\delta\phi_x$		10	20	μrad	Note 2. Over full range
	Y	$\delta\phi_y$		10	20	μrad	
Pitch	X	$\delta\theta_x$		10	20	μrad	Note 2. Over full range
	Y	$\delta\gamma_y$		10	20	μrad	
Roll	X	$\delta\gamma_x$		10	20	μrad	Note 2. Over full range.
	Y	$\delta\theta_y$		10	20	μrad	
Orthogonality error	XY	$\delta\phi_{\mathrm{orth}}$			2	mrad	Note 2
Notes:	1		Design value. Sample tested.				
	2		Measured on test for each unit. Values supplied to user.				

Table 7.1. Sample spec. sheet

Static Characteristics

Or at least slowly moving.

Material

This is the stuff of which most of the stage is made: the properties of various materials have been discussed in Chapter 4, Materials.

The materials thermal expansion coefficient will govern parasitic motions due to temperature changes and its density will influence the mass of the moving part of the stage and thus its resonant frequency. Its Young's modulus will affect the hinge design. Thermal expansion needs to be considered by the user (see *Material thermal expansion coefficient* below) but the other effects are reflected in the stiffness, load and resonant frequency specifications.

Size

The overall outer dimensions of the stage. Particular details, like an aperture in the moving part of the stage, are given in the Comments column. The dimensions are part of the design and are not measured on all units.

Material thermal expansion coefficient

This is the number quoted by our material suppliers for the material of which the stage is made. It is not verified directly by test.

The materials thermal expansion coefficient will determine the movement of a particular point on the stage with respect to temperature and its thermal conductivity will determine how fast it will reach equilibrium after a temperature change. Motion due to temperature changes will be along a line between the point in question on the stage and the isostatic equilibrium point, which is arranged to be at the centre of the stage and will be greatest the further one is away from the equilibrium point (see *metrology frame and kinematic mounting* above). There is no motion at the equilibrium point. In general for high temperature stability choose a material with low thermal expansion coefficient, but be careful: sometimes it is better to have the thermal properties of the stage material the same as the rest of system in which the stage is being used, see the atomic force microscope example in Chapter 4.

Range

This is the maximum closed-loop displacement capability of the stage. The open-loop capability will be somewhat higher as some headroom must be left to deal with the effects of piezo hysteresis and thermal drift. Correct closed loop operation is indicated by the controller's 'ready' flag being true. This is used to measure the closed-loop range.

1. Starting from zero, scan the command in a negative direction in steps of 0.5 µm. Stop when the 'ready' flag remains false after a step. Note that it will flicker false while the stage is moving: ignore this. The command applied before the one that caused the false indication is the maximum negative extent of the range.

2. Repeat the process starting from zero and scanning in a positive direction. This gives the maximum positive extent of the range.

Scale factor

The scale factor is the b_1 mapping coefficient discussed in Chapter 2, Accuracy, Trueness and Precision, *Position linearity and mapping*. As the command units are micrometres, b_1 should be unity: that is if you ask the stage to move one micrometre, it should move one micrometre.

The scale factor is set and verified using a laser interferometer as a measure of true position, see the Appendix to this chapter for a description of the interferometer and scale factor setting. Once set, the scale factor is verified as follows:

1. Mount the interferometer retro-reflector (M2) on the stage, on the axis being verified.

2. Set the command to zero and zero the interferometer readout.

3. Starting from the negative end of the range, step the stage to the positive end of the range ('up') and then back down to the negative end ('down'), taking around 100 steps in each direction and recording the interferometer reading at each step.

4. Each command step now has two position points: an up and a down. Generate an average data set by averaging the two points at each command. Fit a straight line to the average interferometer reading (x_p) v. command (x_c) and derive the slope and error on the slope.

The slope found at step 4 above is the scale factor. It should be unity within the error also determined. The scale factor verification is carried out on all axes of the stage on test.

Scale factor error

This is a 'root of the sum of the squares' (RSS) combination of the accuracy of the interferometer measurement system referred to the definition of the metre, and the error on the slope determined in *Scale factor*, step 4 above

$$\delta b_{x1} = \sqrt{\delta b_{x1.\text{slope}}^2 + \delta x_{m.\text{interf}}^2}$$ (7.2).

A similar equation exists for the other axes.

Static stiffness

This is the ratio of an external force applied to the axis in question, to the displacement that force creates, with the control loop switched off. It is relevant only for computing the resonant frequency and change in range if the stage is used to push against a spring force: see Chapter 6, Piezos, *Static performance*. It should be noted that the closed-loop stiffness is effectively infinite for slowly applied external forces.

The stiffness is not measured directly on every unit but can be inferred as being in specification if the lowest resonant frequency is correct as we have

$$f_0 = \frac{1}{2\pi} \sqrt{\frac{k_x}{m_{\text{load}} + m_{\text{stage}}}} \qquad (7.3).$$

Allowable load

This is the largest mass that can be attached to the stage and have it still control reliably at the slowest speed setting. The weight of this mass will not cause damage.

This parameter is sample tested by ensuring that a stage with the specified maximum load applied controls with the 'slow' bandwidth set: see *Effective bandwidth* below.

Dynamic characteristics

Aside from the 'static' accuracy of motion, the dynamic performance of the system is also important because speed is critical to many applications. Ideally there would be no phase lag between command and position and the mechanism would respond perfectly to a step input i.e. zero rise time, over shoot, and settle time.

The system response depends on the resonant frequency, the damping factor and the PID parameters set. See in Chapter 3, Servo Control for a full discussion of this.

Effective bandwidth

Three sets of PID parameters are supplied with each stage: 'fast', 'medium' and 'slow'. The fast set will give the shortest settling time possible, with a load mass of 0 to 20 g. The medium set will give specified operation with loads up to 100 g. The slow set will give stable operation with all specified loads. Note that the PID parameters can be adjusted by the user for a specific set-up: one is not limited to fast, medium or slow!

The effective bandwidth is defined as the frequency at which the displacement of the stage is 3 dB down on the displacement at dc for a sinusoidaly varying command. The sensor circuitry built into the controller always works at its maximum bandwidth and this measured position information is available to the user. It is thus convenient to use the system's own sensors to measure the time and frequency response characteristics using the 'snapshot' operation mode of the controller. This mode can be set to apply a command impulse and record in the controllers own memory up to about 0.65 s of measured position data, sampling every 40 µs. The data can then be down-loaded to the host computer and Fourier transformed to get the frequency response.

1. Load the PID parameters for the speed of interest.

2. Command an impulse response and down-load the response data.

3. Perform a Fourier transform on the data to get the frequency response plot (Excel or MatLab can do this).

4. Read off the frequency at which the response is 3 dB down on that at dc and compare it with the specification.

Fig. 7.3 shows a typical pair of response curves (medium PID) with a 3 dB response at 190Hz.

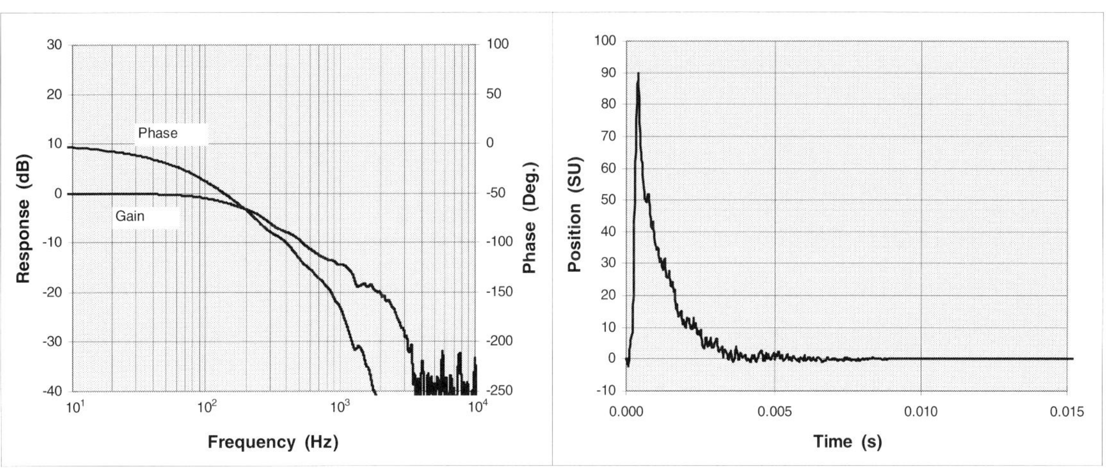

Fig.7.3. Impulse and frequency response curves. (NPS-Z-15B, closed loop, medium PID)

Lowest Resonant frequency

For a linear, second-order, damping-free mechanical system, the resonant frequency is determined by system stiffness and mass. In an optimally designed mechanism, the stiffness is usually dominated by the stiffness of piezo stacks in its translation axis see Chapter 6, Piezos: Dynamic performance. For a stage with motion amplification, the effective stiffness of the piezo actuator will be reduced as shown by equation 7.1.

Reducing the mass will increase the system resonant frequency. However, as the mass of the platform decreases the stage performance becomes more sensitive to the influence of the load mass, i.e. the resonant frequency will drop down rapidly as the mass of the specimen increases.

The easiest way to verify the lowest resonant frequency is to apply an impulse to the stage in open loop mode. The stage will then ring at its natural frequency. Again the controller snapshot mode is used.

1. Attach the required test mass to the stage, set open loop mode and command an impulse response.

2. Fourier transform the impulse response data. The lowest peak is the resonant frequency of interest. Fig. 7.4 shows a ringing response with a peak at approximately 2.2 kHz.

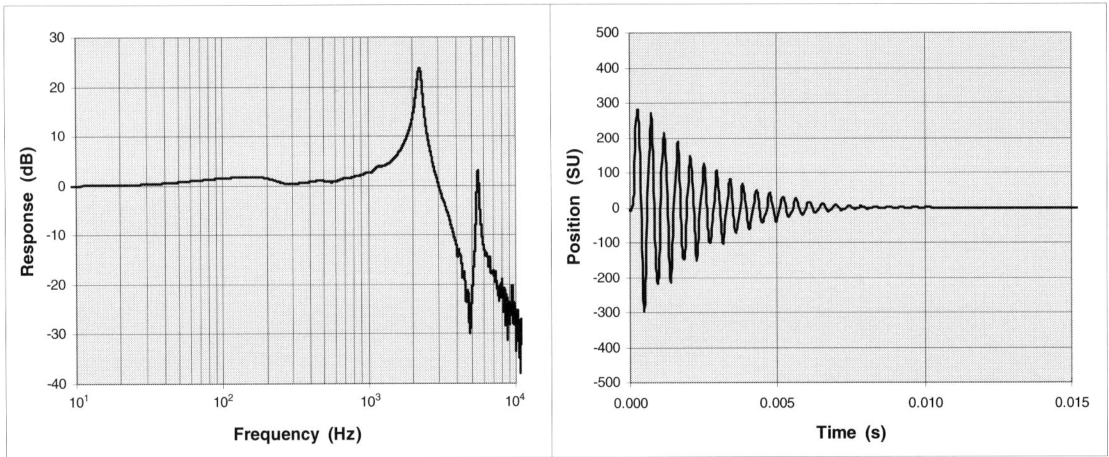

Fig. 7.4. Ringing response (NPS-Z-15B, open loop)

Small signal settling time

The settling time for a small signal is governed primarily by the servo loop response characteristics which in turn are set via the PID parameters. This is in contrast to the settling time for a large signal which will be dominated by the slew-rate. See Chapter 3, Servo Control for a discussion of small and large signals. In general 0.5 μm or less is a small signal.

The measurement technique again uses the 'snapshot' mode except that a step command is used rather than an impulse as was used for bandwidth measurement.

1. Attach a test mass to the stage and load the appropriate set of PID parameters.

2. Command a snapshot with a 0.5 μm step or less.

3. Plot the step response and read off the time taken for the measured position to get to and stay to within ±2 % of the final value.

Fig. 7.5 shows typical small and large signal responses.

Large signal settling time

This measurement is the same as the small signal settling time, only a large step is commanded

1. Attach a test mass to the stage and load the appropriate set of PID parameters.

2. Command a snapshot with a step of 80 % of full range.

3. Plot the step response and read off the time taken for the measured position to get to and stay within 2 % of the final value.

Fig. 7.5 shows typical small and large signal responses.

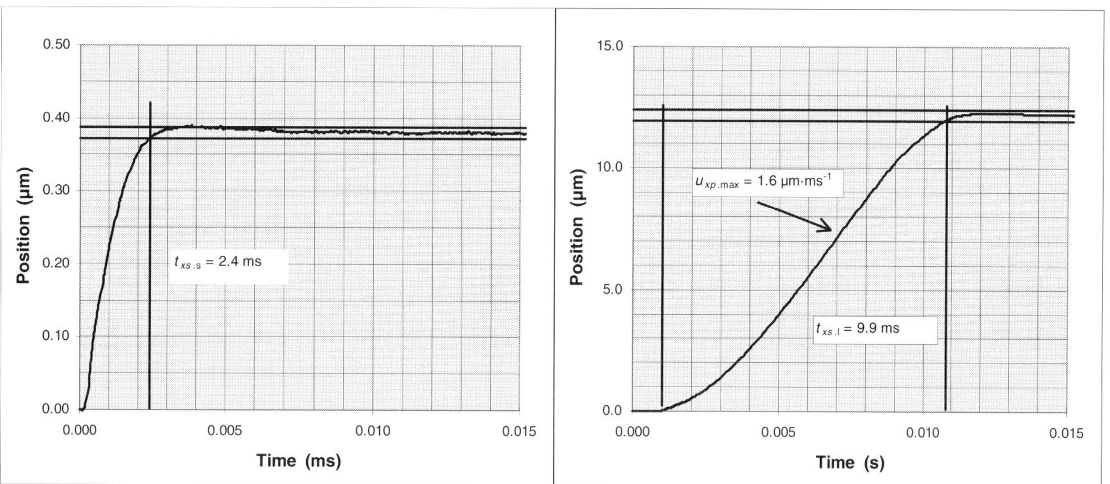

Fig. 7.5. Typical small and large signal response plots. (NPS-Z15-B, medium PID)

Slew rate

This is the slope of the linear section of the large signal response measured above: see Fig. 7.5. It depends on the current drive capability of the piezo drive amplifiers and the capacitance of the piezo actuators, though this must not be allowed to dominate. In practice the integrator slew rate is set to be a bit less than this and govern the slew rate. See Chapter 3, Servo Control for a full discussion.

Position noise

Again this uses the snapshot mode, this time with no command. Position noise in closed loop mode is dominated by the internal sensor noise, so that is measured.

1. Freeze the controller output.

2. Take a snapshot with no command. This gives measurement noise with a small contribution from the piezo drive amplifiers and external vibration.

3. Compute the standard deviation of the measured position data set. This is the measurement noise, $\delta x_{m.n}$ over the full measurement bandwidth B_{xm}.

4. Derive the position noise within the previously measured bandwidth:

$$\delta x_{p.n} = \delta x_{m.n} \sqrt{\frac{B_{xp}}{B_{xm}}} \qquad (7.4).$$

This is not exactly correct as it contains the piezo drive noise $\delta x_{p.ndrive}$, which will be partially servoed out in closed loop mode. However that component is usually very small compared with measurement noise, except when the mechanical amplification is very large (>10).

Mechanical error terms

Hysteresis

Motion along a measurement axis will be free of hysteresis as the capacitance sensors are intrinsically hysteresis free. Parasitic motions however by definition are not directly measurable by the sensors. They may therefore exhibit hysteresis if there is any friction in the system or if the motion axis is not co-linear with the measurement axis (cosine error). In the latter case the component of motion orthogonal to the measurement axis caused by the piezo will exhibit the hysteresis of the piezo. Obviously this must be minimised by design.

In Chapter 6, Piezos, we showed typical piezo hysteresis loops where the difference between the 'up' and 'down' directions is very distinct. In a closed loop stage the difference is very small and indeed quite often masked by noise. Some smoothing is required to show up hysteresis rather than the noise. Hysteresis is measured using the data taken during *Scale factor* verification on page 87. During that test the stage was scanned from the negative end of the range to the positive end and then from the positive end to the negative end. We will call these scans the 'up' and 'down' scans respectively.

1. Generate a plot of hysteresis over the range by subtracting the 'down' scan from the 'up' scan

$$\text{Hyst} = x_{p.\text{up}} - x_{p.\text{down}} \qquad (7.5)$$

and express this as a percentage of the full scan range.

2. Ideally this will just show noise about the X axis. Any hysteresis will show as a large scale structure in this line. Backlash would show as an overall shift up or down.

3. Smooth the hysteresis curve to average out noise.

4. Quote the hysteresis as the maximum distance of the smoothed curve from the X axis.

This sounds rather complicated but Fig. 7.6 should help!

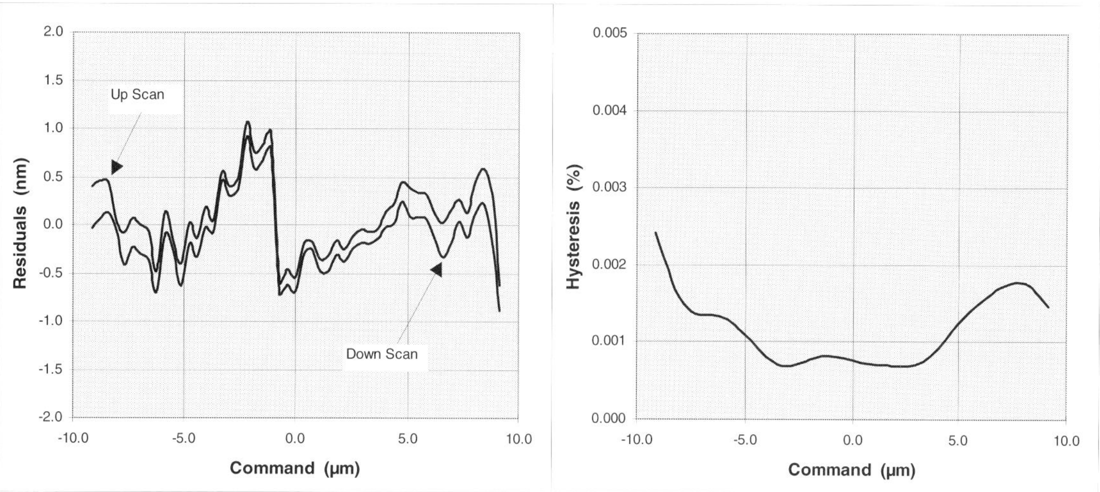

Fig. 7.6 a) Residuals to 'up' and 'down' scans   *b) Smoothed difference, i.e. hysteresis*

This shows a plot of up and down *residuals* to the straight line fit (it is easier to see the errors on residual plots rather than the actual scans) and the smoothed difference between the two, expressed as a percentage of the full range. This is the hysteresis curve and is arithmetically the same as equation 7.5. The maximum hysteresis is just under 0.0025 % for this stage, which is a NPS-Z15-B. In this example there is a component that is constant over the range indicating some backlash (about 0.001 % or 0.15 nm)

Non-linearity

Non-linearities are caused by geometrical effects in the capacitance sensors, see Chapter 5, Capacitance sensors. As the system on test will have been linearised to fourth order during the calibration procedure, any remaining non-linearities are likely to be of higher order and again they will be small and perhaps masked by noise

Again smoothing techniques must be used if we want to get a measure of non-linearity. The scan data used for *Scale factor* verification above is used.

1. Generate a residual curve by subtracting the best fit straight line from the averaged data

$$\mathrm{Res}_{avg} = x_p - \left(b_0 + b_1.x_c\right) \tag{7.6}.$$

Ideally this will just show noise around the *X* axis. Any large scale structure to the curve indicates non-linearity.

2. Smooth the residual curve to average out noise.

3. Quote the non-linearity as half the peak to peak value of this curve.

Fig. 7.7 shows the idea.

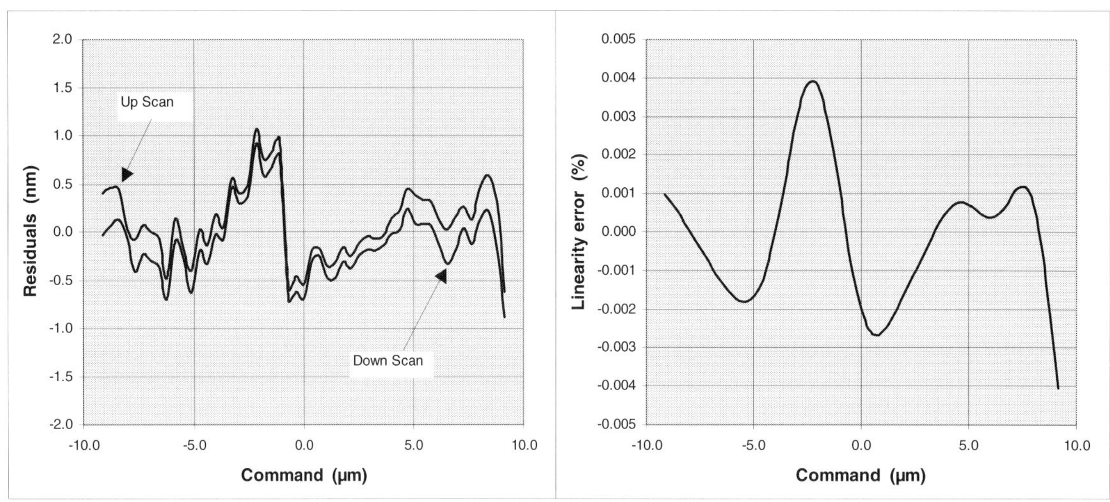

Fig. 7.7 a) Residuals to 'up' and 'down' scans   *b) Smoothed average, i.e. linearity error plot*

This shows the same plot of up and down residuals that was used for hysteresis determination. The linearity error plot is the smoothed *average* of these two, expressed as a percentage of the range, rather than the difference. This is arithmetically the same as equation 7.6. The non-linearity quoted is half the peak to peak value of this curve, or around 0.004 %. Looking at Fig. 7.7b, it would seem that a fifth order mapping polynomial would reduce the error still further without hitting the noise limit. See *Improving performance* later in this chapter.

Roll, pitch and yaw

All stages are designed to produce 'pure' motions, that is if an X-Y stage is commanded to move in *X* it will only move in *X* and not in *Y*, *Z*, γ, θ, ϕ as well. Unfortunately there will be manufacturing errors in the spring strip mechanisms that provide the guided motion: see *Guided motion* earlier in this chapter. These will cause unwanted translations and rotations. **WARNING:** Read the definitions of roll, pitch and yaw in Chapter 2 before attempting to understand these measurements!

Roll, pitch and yaw are minimised by design and are measured using a 2D autocollimator and flat mirror mounted on the stage, see Figs. 7.8 and 7.9. Fig. 7.8 is used to measure pitch and yaw for *X* and roll and pitch for *Y*. Fig 7.9 is used to measure pitch and yaw for *Y* and roll and yaw for *X*. Note that *X* yaw and *Y* pitch can be measured with either or both setups. For a single axis stage, ignore the references to *Y*.

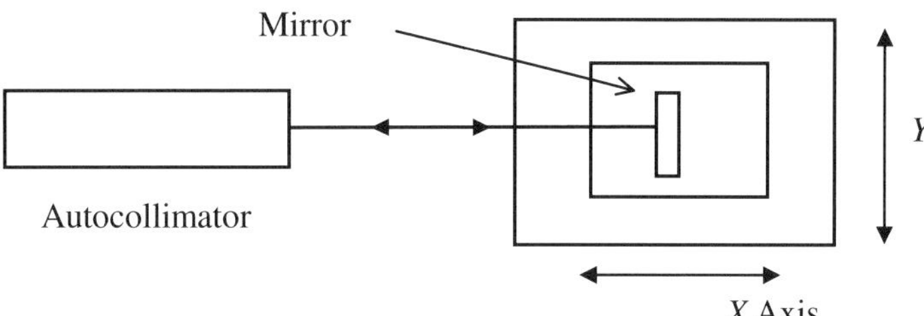

Fig.7.8. X pitch and yaw; Y roll

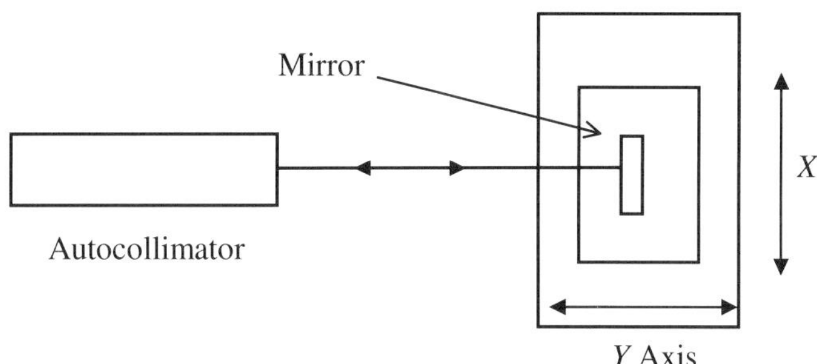

Fig.7.9. Y pitch and yaw; X roll

The autocollimator has an angular resolution of 0.1 µrad.

For each set-up, command one axis to zero and scan the other over its full range. The change of angles as read by the autocollimator in the plane of the paper and at right angles to it, as the relevant axis is scanned, gives roll pitch and yaw information as shown in Table 7.2.

Set-up	Scan *X*		Scan *Y*	
	In plane	**Out of plane**	**In plane**	**Out of plane**
Fig.7.8	*X* yaw, $\delta\phi_x$	*X* pitch, $\delta\theta_x$	*Y* pitch, $\delta\phi_y$	*Y* roll, $\delta\theta_y$
Fig.7.9	*X* yaw, $\delta\phi_x$	*X* roll, $\delta\gamma_x$	*Y* pitch, $\delta\phi_y$	*Y* yaw, $\delta\gamma_y$

Table 7.2. Roll, pitch and yaw determination

Orthogonality and crosstalk

Crosstalk is an unwanted motion along or about a given axis caused by a wanted motion in another. Roll pitch and yaw are examples of crosstalk, but here we are more interested in, say, unwanted x displacements produced by wanted y displacements. In this case the crosstalk is also known as 'runout'. Crosstalk is mainly determined by manufacturing tolerances. If the axes of the two pairs of sensors in the X-Y stage are not orthogonal, then a motion in X will produce a parasitic motion in Y and vice versa. The departure from orthogonality of the axes is expressed as an angular deviation from 90° ($\pi/2$ rad). Using modern manufacturing technology, the orthogonality of the sensor axes can be generally controlled to within 0.5 mrad which gives a crosstalk of 0.5 nm·µm^{-1} (i.e. 0.05%) in the *X-Y* plane.

It is important not to confuse crosstalk produced by non-orthogonality with the effects of yaw. If one is concerned about a point away from the coordinate system origin (the centre of the moving part of the stage in an X-Y system), then yaw will cause an off-axis motion. Roll, pitch and yaw are considered to be rotations about the origin: crosstalk due to non-orthogonality of the axes causes an off-axis motion of a point *at the origin*. At the origin of an X-Y stage, there will be no off-axis motion due to yaw, to first order.

Orthogonality is measured directly using the laser interferometer using the set-ups of Fig. 7.10 a and b. The stage is mounted on a precision

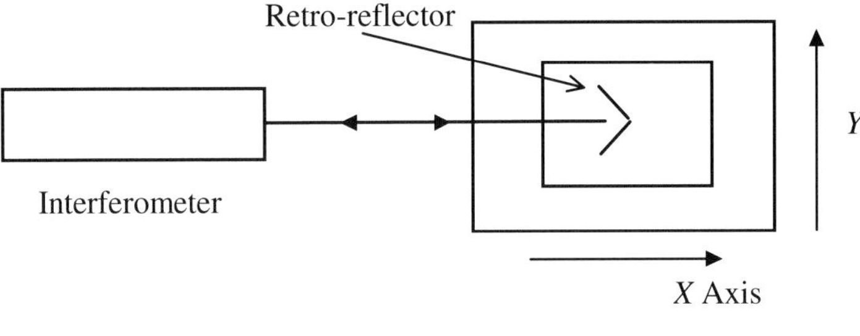

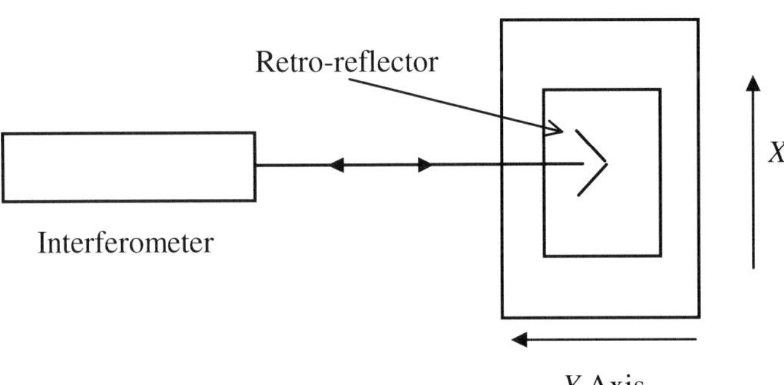

Fig. 7.10 Orthogonality determination

table that can be rotated through 90° with an accuracy of 50 μrad.

In these figures the laser, beamsplitter and stationary retro-reflector are contained in the box labeled 'interferometer'.

1. Considering Fig. 7.10a, mount the retro-reflector on the moving part of the stage with its apex at the origin. The effective measurement point of a laser interferometer is the retro-reflector apex.

2. Command the stage to zero in X and scan over the full range in Y recording any motion registered by the interferometer. If the Y axis of the stage is at right angles to the measurement laser beam, no motion will be recorded. It is unlikely to be exactly lined up so there will be a displacement. Calculate the misalignment angle

$$\phi_1 = \frac{\mathrm{d}x}{\mathrm{d}y} \tag{7.7}$$

where dx is the motion along the x axis produced by the Y motion dy. Remember to observe the sign of the motions.

3. Turn the stage through 90° *anti-clockwise* and re-position the retro-reflector to give the set-up of Fig. 7.10b.

4. Command the stage to zero in Y and scan over the full range in X recording any motion registered by the interferometer. Calculate the misalignment angle

$$\phi_2 = \frac{\mathrm{d}y}{\mathrm{d}x} \tag{7.8}$$

where dy is the motion along the y axis produced by the X motion dx. Remember to observe the sign of the motions.

5. If the axes are orthogonal, then the two angles measured above will be the same. The orthogonality is defined

$$\delta\phi_{\mathrm{orth}} = \phi_2 - \phi_1 \tag{7.9}$$

and the actual angle in radians between the X and Y axes will be

$$\phi_{XY} = \frac{\pi}{2} + \delta\phi_{\mathrm{orth}} \tag{7.10}.$$

USING NANOMECHANISMS

Each user will have a specific set of requirements, but there are some general principles that can be considered when choosing and using NanoMechanisms.

Configuring multi axis systems

Abbe error

When specimens are mounted on to the NanoMechanism, Abbe errors have to be considered carefully due to the parasitic angular motions. Small angular errors can have a large affect at the nanometre level: for example, a tilt of just 1 µrad with an offset of 1 mm gives a 1 nm position error. To reduce this effect, specimens should be positioned as close as possible to the measuring axes of the sensors. For example, in an XYZ 3 axis NanoMechanism system the specimen holder is located at the point which is co-linear with the sensor measuring axes, as shown in figure 7.11 The effects of yaw, pitch and roll of the XY stage can thus be minimised.

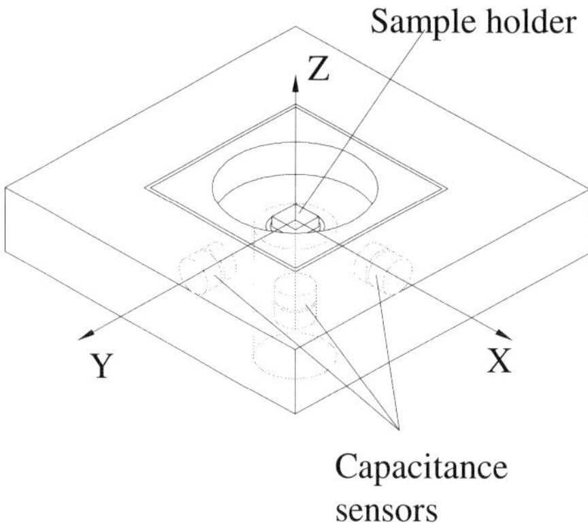

Fig. 7.11 Minimising Abbe error

Cosine error

Take care to align the motion axes of the stage with the axes of the system in which it is used. If there is an angular offset, then the users system may see less motion than is actually occurring: see Chapter 2, Accuracy, Trueness and Precision, *Cosine and Abbe errors*.

Customised configurations

It is not necessary to buy a stage to get closed loop multi-axis control. Many systems benefit from using separate translators and sensors in specially constructed mechanisms. This can enable the sensors to be placed as close as possible to the point where control is required. Such systems are by their nature very specific to individual requirements but there are five basic questions that should be asked when choosing the positioning and measuring components:

1. How far?	Determines the technology; <100 µm: piezo, >100 µm: mechanical amplification of piezo motion or DC motors driving screws. See *Mechanical amplification* above.
2. How fast?	Determines the power rating of the driving amplifier: See Chapter 3, Servo control
3. How heavy?	Determines the mechanical resonant frequency (via the stiffness) which in turn limits the closed loop bandwidth. See Chapter 3, Servo control.
4. High linearity?	Determines the choice of sensor size and the measuring range. See Chapter 5, Capacitance sensors.
5. Low noise?	Determines the closed loop bandwidth and the measuring range. See Chapter 5, Capacitance sensors.

Improving performance

Stages and controllers are supplied optimised for most general applications but there are some tricks that can be applied to squeeze the last drop of performance out of a system.

Mapping correction

An obvious place to start is with mapping correction. During non-linearity determination (see *Static parameters: Non-linearity* above) a sixth order polynomial is fitted to the residual curve. The reverse polynomial can be used to map out the higher order terms by applying it to the command before sending it to the controller. Use the direct polynomial to correct the position read-back.

Abbe and cosine error correction

The roll, pitch, yaw and orthogonality are measured for each stage, so their effects can be compensated for a given point on the stage by calculating the X and Y commands required to position the required point.

Command filtering

Dynamic performance can also often be improved by not exciting resonances in the first place. Convolve Inc. have some marvelous techniques for modifying the command so as not to excite resonances and thus improve settling times. This works well if one is interested in controlling a somewhat resonant system attached to a stage.

Temperature control

All points on the stage will move with respect to the isostatic mounting datum (origin) due to thermal expansion of the material from which the metrology stage is manufactured. Even for a SuperInvar stage of size 100 x 100 mm, a 1 K temperature change will cause 30 nm change in dimension ($\alpha = 0.3$ ppm·K^{-1}). For high accuracy positioning over a period of time, temperature control is recommended.

APPENDIX. TRUE POSITION MEASUREMENT

Back in Chapter 2 we introduced the concept of a true position. Calibration consists of measuring the true position of a stage with respect to the commanded position. It is measured as the displacement in all six degrees of freedom for a given command from the position defined by command zero. But what is our true metre rule?

The second is defined as the duration of 9 192 631 770 periods of the radiation corresponding to the transition between the two hyperfine levels of the ground state of the caesium-133 atom. The metre is then defined as the length of path traveled by light in vacuum during a time interval of 1/299 792 458 of a second. So to get a perfect metre to which we can refer, all we have to do is get a few caesium-133 atoms and see how far a light beam goes in 9 192 631 770/299 792 458 periods of the above mentioned hyperfine transition. In a perfect vacuum.

The above may be a bit tongue-in-cheek, but it does illustrate that at least there is an absolute definition of the quantity we want to measure: distance (as long as we can ignore Einstein!). Doing it directly as suggested above, however, is a little impractical and usually one uses a more convenient metre rule. The He-Ne laser provides this.

The wavelength of the He-Ne laser line is nominally 632.817 3 nm in vacuum, which can be corrected for use in air, but it will vary a bit depending on the mode structure of the particular laser and atmospheric fluctuations. The accuracy will be at least one part in 10^7 which is 0.01 nm in 100 µm. The He-Ne laser is used in a Michelson laser interferometer to measure displacement.

The Michelson interferometer

Fig. 7.12 is a schematic diagram of a Michelson interferometer.

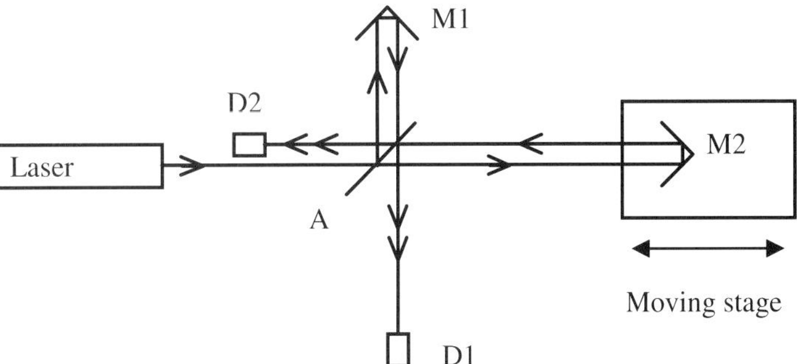

Fig. 7.12. Michelson interferometer

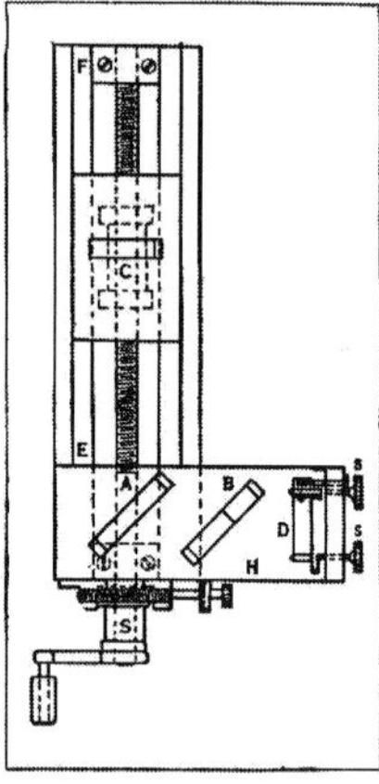

An early Michelson interferometer

Light from the laser hits beam splitter A. Half of it is transmitted to the retro-reflector M2 mounted on the stage on test and the other half is reflected to the fixed retro-reflector M1. The two beams return to the beam splitter where they superimpose and fall on detectors D1 and D2. As the stage on test moves, alternate constructive and destructive interference will take place between the beams falling on detector D1. Similar interference will take place between the beams falling on D2 but there will be a phase difference between these and those on D1. This happens because light reflected from the beam splitter will be phase shifted with respect to the light transmitted.

A bright 'fringe' is seen every half wavelength ($\lambda/2$) of motion of the stage, so this provides a measurement scale with tick-marks every 316.408 65 nm. Electronic interpolation of the D1 and D2 signals can give a readout with a resolution of a couple of nanometres for a general purpose system, or about 0.3 nm for a very precise system designed by the UK National Physical Laboratory and used at Queensgate for some precision work. The relative phase of the signals from D1 and D2 can be used to determine the direction of travel of the stage.

Stage calibration

Setting of the scale factor is the major calibration activity performed on all stages.

A retro-reflector (M2) is mounted on the stage, on the axis being calibrated and the following procedure carried out.

1. Set the command to zero and zero the interferometer readout.

2. Step the stage over its full range in both directions taking around 100 steps in each direction and recording the interferometer reading at each step.

3. Fit a fourth order power series to the interferometer reading (x_p) v. command (x_c) and derive the 'b' coefficients.

4. Load the 'b' coefficients into the controller to correct the stage sensor output, see Chapter 3, Servo Control: Static mapping.

5. Repeat steps 1 to 3 and record the new 'b' coefficients for supply to the user and later use. Note that b_1 should now be unity and the remaining coefficients very small as the system will now be virtually linear.

The mapping function is dictated primarily by the geometry of the capacitance sensors in the stage, see Chapter 5, Capacitance Sensors: Practical capacitors. The controller therefore actually stores the coefficients in an EPROM located in the stage. This means that if the controller is to be used with a variety of stages, the correct mapping functions will always be utilised.

The mapping procedure is carried out on all axes of the stage on test.

REFERENCES

S. T. Smith and D. G. Chetwynd, Foundations of Ultraprecision Mechanism Design, *Gordon and Breach Science Publishers, 1994*

Convolve Inc. 1 Quarter Mile Road, Armonk, NY 10504

TERMS

This is a list of terms used to describe system performance. Description terms in italics refer to other entries in the glossary but note that terms and sub-terms only appear once: if you can't find 'measurement hysteresis' try 'hysteresis, measurement'. In most cases the symbol is given only for the 'x' component of the term in question: y, z, γ, θ, ϕ may be substituted as required.

Term	Symbol	Description	Units	Page(s)
Abbe error		After *Ernst Karl Abbe (1840-1905)*, a German mathematician and physicist. A measurement error produced by *roll, pitch* or *yaw* when the point of interest is not on the measurement axis..		
measurement	$\delta x_{m\theta.abbe}$	Measurement error along the X axis caused by a rotation about the Y axis coupled with a Z *Abbe offset*.	nm	12, 13, 24, 98
	$\delta x_{m\phi.abbe}$	Measurement error along the X axis caused by a rotation about the Z axis coupled with a Y *Abbe offset*.		
	$\delta y_{m\gamma.abbe}$	Measurement error along the Y axis caused by a rotation about the X axis coupled with a Z *Abbe offset*.		
	$\delta y_{m\phi.abbe}$	Measurement error along the Y axis caused by a rotation about the Z axis coupled with a X *Abbe offset*.		
	$\delta y_{m\gamma.abbe}$	Measurement error along the Z axis caused by a rotation about the X axis coupled with a Y *Abbe offset*.		
	$\delta y_{m\theta.abbe}$	Measurement error along the Z axis caused by a rotation about the Y axis coupled with a X *Abbe offset*.		
position		A position error produced by *roll, pitch* or *yaw* when the point of interest is not on the motion axis..		
	$\delta x_{p\theta.abbe}$	Position error along the X axis caused by a rotation about the Y axis coupled with a Z *Abbe offset*.	nm	12, 13, 24, 98
	$\delta x_{p\phi.abbe}$	Position error along the X axis caused by a rotation about the Z axis coupled with a Y *Abbe offset*.		
	$\delta y_{p\gamma.abbe}$	Position error along the Y axis caused by a rotation about the X axis coupled with a Z *Abbe offset*.		
	$\delta y_{p\phi.abbe}$	Position error along the Y axis caused by a rotation about the Z axis coupled with a X *Abbe offset*.		
	$\delta y_{p\gamma.abbe}$	Position error along the Z axis caused by a rotation about the X axis coupled with a Y *Abbe offset*		
	$\delta y_{p\theta.abbe}$	Position error along the Z axis caused by a rotation about the Y axis coupled with a X *Abbe offset*.		
Abbe offset	$d_{x.abbe}$ $d_{y.abbe}$ $d_{z.abbe}$	Distance of the point of interest from the motion axis.	µm	13, 23
Accuracy				
measurement	δx_{mA}	Arithmetic sum of *measurement trueness* and *precision*, e.g. $\delta x_{mA}=\delta x_{mS}+\delta x_{mR}$	nm	7
position	δx_{pA}	Arithmetic sum of *position trueness* and *precision*, e.g. $\delta x_{pA}=\delta x_{pS}+\delta x_{pR}$	nm	17
Axes	X,Y,Z γ,θ,ϕ	A reference frame used for distance, motion and rotation measurement. Queensgate uses a right-handed orthogonal Cartesian system. The origin of the reference frame is the *true position* when the *position commands* x_c, y_c, z_c are zero.	-	5,17
Backlash	-	*Hysteresis* that is constant over the range	%	22

Term	Symbol	Description	Units	Page(s)
Bandwidth				
measurement	B_{xm}	The frequency range over which measurements are taken: dc to 3dB down point	Hz	14, 66
positioning	B_{xp}	The frequency range to which the stage can respond: dc to 3dB down point	Hz	21
Cosine error				
measurement		A measurement error produced when the measurement axis is misaligned with respect to the motion axis.		
	$\delta x_{m.\cos\phi}$	X measurement error produced by misalignment of the X axes about the Z axis.	nm	12, 98
	$\delta x_{m.\cos\theta}$	X measurement error produced by misalignment of the X axes about the Y axis.		
	$\delta y_{m.\cos\gamma}$	Y measurement error produced by misalignment of the Y axes about the X axis.		
	$\delta y_{m.\cos\phi}$	Y measurement error produced by misalignment of the Y axes about the Z axis.		
	$\delta z_{m.\cos\gamma}$	Z measurement error produced by misalignment of the Z axes about the X axis.		
	$\delta z_{m.\cos\theta}$	Z measurement error produced by misalignment of the Z axes about the Y axis.		
position		A position error produced when the motion axis is misaligned with respect to the true position measuring device axis.		
	$\delta x_{p.\cos\phi}$	X position error produced by misalignment of the X axes about the Z axis.	nm	12, 98
	$\delta x_{p.\cos\theta}$	X position error produced by misalignment of the X axes about the Y axis.		
	$\delta y_{p.\cos\gamma}$	Y position error produced by misalignment of the Y axes about the X axis.		
	$\delta y_{p.\cos\phi}$	Y position error produced by misalignment of the Y axes about the Z axis.		
	$\delta z_{p.\cos\gamma}$	Z position error produced by misalignment of the Z axes about the X axis.		
	$\delta z_{p.\cos\theta}$	Z position error produced by misalignment of the Z axes about the Y axis.		
Crosstalk		A measurement error along a given axis due to motion along another axis. See also *orthogonality* and *Abbe error*.		
measurement	δx_{my}	X measurement caused by a displacement along Y	nm	
	δx_{mz}	X measurement caused by a displacement along Z		
	δy_{mx}	Y measurement caused by a displacement along X		
	δy_{mz}	Y measurement caused by a displacement along Z		
	δz_{mx}	Z measurement caused by a displacement along X		
	δz_{my}	Z measurement caused by a displacement along Y		
position		A position error along a given axis due to motion along another axis. See also *orthogonality* and *Abbe error*. Also known as 'runout'		
	δx_{py}	Displacement along X axis caused by a displacement along Y	nm	
	δx_{pz}	Displacement along X axis caused by a displacement along Z		
	δy_{px}	Displacement along Y axis caused by a displacement along X		
	δy_{pz}	Displacement along Y axis caused by a displacement along Z		
	δz_{px}	Displacement along Z axis caused by a displacement along X		
	δz_{py}	Displacement along Z axis caused by a displacement along Y		
Dead band				
measurement	$\delta x_{m.\text{dead}}$	A region of *true positions* over which there is no change in *measured position*.	nm	22
position	$\delta x_{p.\text{dead}}$	A region of *desired positions* over which there is no change in *true position*.	nm	22

Term	Symbol	Description	Units	Page(s)
Displacement	d_x, d_y, d_z	A change in position	nm	5
Drift measurement	$\delta x_{m.\mathrm{drift}}$	A change in mean *measured position* on a time scale of minutes or more with the *true position* constant. It includes the effect of *measurement temperature coefficient* and other environmental effects.	nm	17
position	$\delta x_{p.\mathrm{drift}}$	A change in mean *true position* on a time scale of minutes or more with the *commanded position* constant. It includes the effect of *position temperature coefficient* and other environmental effects.	nm	17
Hysteresis measurement	$\delta x_{m.\mathrm{hyst}}$	The difference between *measured positions* when a single *true position* is approached from both ends of the range.	%	16
position	$\delta x_{p.\mathrm{hyst}}$	The difference between *true positions* when a single *desired position* is approached from both ends of the range. The maximum value is quoted, as a percentage of the range.	%	22-24, 75,76,82
Interference	–	General term for unwanted noise contributions from external sources such as the line voltage, processor clocks, radio transmitters etc.	pm	14, 67
Linearity error measurement	$\delta x_{m.\mathrm{lin}}$	*measurement mapping error* with a first order power series (straight line) *measurement mapping function*.	%	8
position	$\delta x_{p.\mathrm{lin}}$	*Position mapping error* with a first order power series (straight line) *position mapping function*.	%	19
Mapping accuracy measurement	δa_{x0}, δa_{x1}, δa_{x2}, δa_{x3}, δa_{x4}	A measure of how well *the measurement mapping polynomial* describes the actual sensor performance. Quoted as errors on the *measurement mapping coefficients*.	µm, %, µm^{-1}, µm^{-2}, µm^{-3}	9
position	δb_{x0}, δb_{x1}, δb_{x2}, δb_{x3}, δb_{x4}	A measure of how well *the position mapping polynomial* describes the actual positioning performance. Quoted as errors on the *position mapping coefficients*.	µm, %, µm^{-1}, µm^{-2}, µm^{-3}	20
Mapping coefficients measurement	a_{x0} a_{x1} a_{x2} a_{x3} a_{x4}	Elements of the 'A' row vectors. See *Mapping polynomial, measurement*..	µm, %, µm^{-1}, µm^{-2}, µm^{-3}	20
position	b_{x0} b_{x1} b_{x2} b_{x3} b_{x4}	Elements of the 'B' row vectors. See *Mapping polynomial, position*.	µm, %, µm^{-1}, µm^{-2}, µm^{-3}	20
Mapping error measurement	$\delta x_{m.\mathrm{map}}$	One half of the maximum peak to peak amplitude of *the residual curve* to the *measurement mapping function*, expressed as a percentage of the full measurement range.	%	9
position	$\delta x_{p.\mathrm{map}}$	One half of the maximum peak to peak amplitude of the *residual curve* to the *position mapping function*, expressed as a percentage of the full positioning range.	%	19

Term	Symbol	Description	Units	Page(s)
Mapping function				
measurement	$x_m{=}\mathrm{f}(x_p)$	A general function that predicts *a position measurement* given an *true position*.	-	8
position	$x_p{=}\mathrm{f}(x_c)$	A general function that predicts an *true position* given a *position command*.	-	20
Mapping polynomial				
measurement	$x_m{=}A_XX_P$	A power series representation of *a measurement mapping function*. The 'A' terms are row vectors *of measurement mapping coefficients* and X_A, Y_A and Z_A are column vectors of powers of the *true position* x_a, y_a, z_a.	-	8
position	$x_p{=}B_XX_C$	A power series representation of *a position mapping function*. The 'B' terms are row vectors *of position mapping coefficients* and X_C, Y_C and Z_C are column vectors of powers of the *command position* x_c, y_c, z_c.	-	20
Noise				
measurement	$\delta x_{m.n}$	The 1σ spread of *position measurements* when the position being measured is stationary. It is a combination of *sensor noise* and *quantisation noise*.	nm	13
position	$\delta x_{p.n}$	The 1σ spread of *true positions* when the position command is stationary.	nm	20
Noise coefficient		A system constant that enables the *measurement noise density* to be determined knowing the capacitor geometry.	$Hz^{-1/2}$	65,66
measurement	$k_{xm.ndens}$			
position	$k_{xp.ndens}$	A system constant that enables the *position noise density* to be determined knowing the capacitor geometry.	$Hz^{-1/2}$	65,66
Noise, Gaussian				
measurement	$\delta x_{m.ng}$	Contribution to *sensor noise*, generally with uniform *density* over the *bandwidth*, caused by thermal effects in electronic components.	nm	14
position	$\delta x_{p.ng}$	Contribution to *position noise*, generally with uniform *density* over the *bandwidth*, caused by thermal effects in electronic components.	nm	14
Noise density				
measurement	$\delta x_{m.ndens}$	The RMS *Gaussian measurement noise* per unit *bandwidth*.	$pm{\cdot}Hz^{-1/2}$	15
position	$\delta x_{p.ndens}$	The RMS *Gaussian position noise* per unit *bandwidth*	$pm{\cdot}Hz^{-1/2}$	20
Noise, piezo drive	$\delta x_{p.ndrive}$	Contribution to position noise from the piezo drive electronics	nm	21
Noise, quantisation				
measurement	$\delta x_{m.nquant}$	Contribution to *measurement noise* due to the finite number of bits used in digitisation	nm	16
position	$\delta x_{p.nquant}$	Contribution to *position noise* due to the finite number of bits used in digitisation	nm	26
Noise, sensor	$\delta x_{m.nsens}$	Contribution to *measurement noise*. A combination of *Gaussian noise* and *interference*	nm	13
Non-linearity	-	See Linearity error		
Orthogonality	-	The property of being at right angles. *Orthogonality error* will cause *crosstalk*.		96
Orthogonality error	$\delta\phi_{orth}$ $\delta\gamma_{orth}$ $\delta\theta_{orth}$	The departure of the angle between the *X* and *Y* axes from 90°. The departure of the angle between the *Y* and *Z* axes from 90°. The departure of the angle between the *Z* and *X* axes from 90°.	mrad	96
Pitch	$\delta\theta_x,$ $\delta\phi_y,$ $\delta\gamma_z$	A rotation about the: *Y* axis when moving along the *X* axis; *Z* axis when moving along the *Y* axis; *X* axis when moving along the *Z* axis.	$\mu rad{\cdot}\mu m^{-1}$	6,95

Term		Symbol	Description	Units	Page(s)
Position		x,y,z	A point in space measured with respect to the *axes*		5
	desired	x_d,y_d,z_d	The *position* that the user requires	μm	20
	command	x_c,y_c,z_c	The *position* set-point sent to a controller	μm	18
	measured	x_m,y_m,z_m	A mean *position* as measured by the system under discussion	μm	7
	true	x_p,y_p,z_p	A *position* as measured with a perfect measurement system	μm	7,18,20
Precision	command	δx_{cR}	The 1σ spread of commands relating to a given *desired position*. Normally this will be the same as the *quantisation noise* caused by digitisation of the *desired position*.	nm	16, 26
	measurement	δx_{mR}	The 1σ spread of *measured positions* including *measurement mapping error, resolution, noise*, and *reproducibility* with the *true position* held constant.	nm	7
	measured position	δx_{mpR}	The 1σ spread of *measured positions* including *measurement precision* and *position precision*	nm	18
	position	δx_{pR}	The 1σ spread of *true positions* including *position mapping error, resolution, noise*, and *reproducibility* with the *commanded position* held constant.	nm	18
Range	measurement	$d_{xp.\max}$	The maximum *displacement* that can be measured.	μm	
	position	$d_{xp.\max}$	The maximum *displacement* that can be generated.	μm	66,82,86
Repeatability	measurement	$\delta x_{m.rpt}$	The 1σ spread of the *measured positions* when the same true half range *displacement* is applied to the sensor repeatedly, under the same operating conditions and in the same direction. Repeatability does not include *hysteresis* or *drift*.	nm	16
	position	$\delta x_{p.rpt}$	The 1σ spread of the *true positions* when the same true half range *displacement* is commanded repeatedly, under the same operating conditions and in the same direction. Repeatability does not include *hysteresis* or *drift*.	nm	24
Reproducibility	measurement	$\delta x_{m.rpd}$	The 1σ spread of the *measured positions* when the same true half range *displacement* is applied to the sensor repeatedly, approaching from both directions. Reproducibility includes *hysteresis, dead band, drift* and *repeatability*.	nm	16
	position	$\delta x_{p.rpd}$	The 1σ spread of the *true positions* when the same true half range *displacement* is commanded repeatedly, approaching from both directions. Reproducibility includes *hysteresis, dead band, drift* and *repeatability*.	nm	22
Resolution	measurement	$\delta x_{m.res}$	The smallest measured *displacement* that can be resolved. This will be the greater of *measurement noise* or *quantisation noise*.	pm	13
	position	$\delta x_{p.res}$	The smallest *displacement* that can be commanded. This will be the greater of *position noise* or *quantisation noise*.	pm	15, 20
Residual curve	measurement		A function of the *true position* and *position measurement* obtained by subtracting the *measurement mapping function* prediction from the *position measurements*. For example: $$\text{Resm}\left(x_m,x_p\right)=x_m-A_x.X_p$$	nm	9

Term	Symbol	Description	Units	Page(s)
Residual curve position		A function of *true position* and *position command* obtained by subtracting the *position mapping function* prediction from the *true position*. For example: $$\mathrm{Resp}(x_p, x_c) = x_p - B_x . X_c$$	nm	9
Resonant frequency lateral	f_{0x}	The frequency of the fundamental lateral oscillation modes along the measurement axes.	Hz	34, 89
rotational	$f_{0\gamma}$	The frequency of the fundamental rotational oscillation modes around the measurement axes.	Hz	34, 89
Roll	$\delta\gamma_x$ $\delta\theta_y$ $\delta\phi_z$	A rotation about the: X axis when moving along the X axis; Y axis when moving along the Y axis; Z axis when moving along the Z axis.	μrad$\cdot\mu$m^{-1}	6,95
Runout		See *crosstalk, position*.		
Scale factor measurement	a_{x1}	The a_1 element of the '**A**' row vectors when a first order power series *mapping function* (straight line) is used. See *Mapping polynomial, sensor*.	–	9
position	b_{x1}	The b_1 element of the '**B**' row vectors when a first order power series *mapping function* (straight line) is used. See *Mapping polynomial, position*.	–	20
system	$SF_{x.sys}$	A multiplier used to convert system units [volts in an analog system or numbers (*system units*: SU) in a digital system] to measurement units (μm)	μm$\cdot$V^{-1} or μm$\cdot$SU^{-1}	
Scale factor error measurement	δa_{x1}	The *sensor mapping accuracy* when a first order polynomial (straight line) *sensor mapping function* is used.	%	9
position	δb_{x1}	The *position mapping accuracy* when a first order polynomial (straight line) *position mapping function* is used.	%	20
Settling time small signal	$t_{xs.s}$	Time taken for the *true position* to get to within 1% of its final value and stay there after a command step of 0.5μm.	ms	37, 41, 90
large signal	$t_{xs.l}$	Time taken for the *true position* to get to within 1% of its final value and stay there after a command step of 80% of the full range.	ms	41, 91
Slew rate	$u_{xp.max}$	The highest rate of change of *true position* with time that can be achieved.	μm$\cdot$ms^{-1}	38, 40, 79, 92
System units	SU	The number units used internally in a digital control system.		
Temperature coefficient measurement	α_{xm}	The rate of change of a *measured position* with respect to temperature with constant *true position*.	nm$\cdot$K^{-1}	43, 86
position	α_{xp}	The rate of change of a *true position* with respect to temperature with constant *commanded position*.	nm$\cdot$K^{-1}	43, 86
Trueness measurement	δ_{xmS}	The difference between the mean *measured position* and the *true position*, x_m-x_p	nm	7
position	δ_{xpS}	The difference between the *true position* and the *desired position*. In a *closed loop* system this will be the same as the *measurement trueness*.	nm	18
Yaw	$\delta\phi_x$ $\delta\gamma_y$ $\delta\theta_z$	A rotation about the: Z axis when moving along the X axis; X axis when moving along the Y axis; Y axis when moving along the Z axis.	μrad$\cdot\mu$m^{-1}	6, 95

SYMBOLS

This lists most of the symbols used in the rest of the text in alphabetical order of the x component. Page numbers refer to the principle definition of the relevant term.

Symbol	Term	Description	Units	Page(s)
a_{x0} a_{x1} a_{x2} a_{x3} a_{x4}	Mapping coefficients, measurement	Elements of the 'A' row vectors. See *Mapping polynomial, measurement.*.	μm, - μm^{-1}, μm^{-2} μm^{-3}	20
a_{x1}	Scale factor, measurement	The a_1 element of the '**A**' row vectors when a first order power series *mapping function* (straight line) is used. See *Mapping polynomial, sensor.*	-	9
α_{xm}	Temperature coefficient, measurement	The rate of change of a *measured position* with respect to temperature with constant *true position*.	$nm \cdot K^{-1}$	43, 86
α_{xp}	Temperature coefficient, position	The rate of change of a *true position* with respect to temperature with constant *commanded position*.	$nm \cdot K^{-1}$	43, 86
b_{x0} b_{x1} b_{x2} b_{x3} b_{x4}	Mapping coefficients, position	Elements of the 'B' row vectors. See *Mapping polynomial, position.*	μm, - μm^{-1}, μm^{-2} μm^{-3}	20
b_{x1}	Scale factor, position	The b_1 element of the '**B**' row vectors when a first order power series *mapping function* (straight line) is used. See *Mapping polynomial, position.*	-	20
B_{xm}	Bandwidth, measurement	The frequency range over which measurements are taken: dc to 3dB down point	Hz	14, 66
B_{xp}	Bandwidth, positioning	The frequency range to which the stage can respond: dc to 3dB down point	Hz	21
δa_{x0}, δa_{x1}, δa_{x2}, δa_{x3}, δa_{x4}	Mapping accuracy, measurement	A measure of how well *the measurement mapping polynomial* describes the actual sensor performance. Quoted as errors on the *measurement mapping coefficients.*	μm, % μm^{-1}, μm^{-2} μm^{-3}	9
δa_{x1}	Scale factor error, measurement	The *sensor mapping accuracy* when a first order polynomial (straight line) *sensor mapping function* is used.	$m \cdot m^{-1}$	9
δb_{x0}, δb_{x1}, δb_{x2}, δb_{x3}, $\ddot b_{x4}$	Mapping accuracy, position	A measure of how well *the position mapping polynomial* describes the actual positioning performance. Quoted as errors on the *position mapping coefficients.*	μm, % μm^{-1}, μm^{-2} μm^{-3}	20
δb_{x1}	Scale factor error, position	The *position mapping accuracy* when a first order polynomial (straight line) *position mapping function* is used.	%	20
$\delta\phi_{orth}$ $\delta\gamma_{orth}$ $\delta\theta_{orth}$	Orthogonality error	The departure of the angle between the X and Y axes from 90°. The departure of the angle between the Y and Z axes from 90°. The departure of the angle between the Z and X axes from 90°.	mrad	96
$\delta\phi_x$ $\delta\gamma_y$ $\delta\theta_z$	Yaw	A rotation about the: Z axis when moving along the X axis; X axis when moving along the Y axis; Y axis when moving along the Z axis.	$\mu rad \cdot \mu m^{-1}$	6, 95
$\delta\gamma_x$ $\delta\theta_y$ $\delta\phi_z$	Roll	A rotation about the: X axis when moving along the X axis; Y axis when moving along the Y axis; Z axis when moving along the Z axis.	$\mu rad \cdot \mu m^{-1}$	6,95

Symbol	Term	Description	Units	Page(s)
$\delta\theta_x,$ $\delta\phi_y,$ $\delta\gamma_z$	Pitch	A rotation about the: Y axis when moving along the X axis; Z axis when moving along the Y axis; X axis when moving along the Z axis.	μrad·μm^{-1}	6,95
d_x, d_y, d_z	Displacement	A change in position	nm	5
$d_{x.abbe}$ $d_{y.abbe}$ $d_{z.abbe}$	Abbe offset	Distance of the point of interest from the motion axis.	μm	13, 23
δx_{cR}	Precision, command	The 1σ spread of commands relating to a given *desired position*. Normally this will be the same as the *quantisation noise* caused by digitisation of the *desired position*.	nm	16, 26
$\delta x_{m.\cos\phi}$ $\delta x_{m.\cos\theta}$ $\delta y_{m.\cos\gamma}$ $\delta y_{m.\cos\phi}$ $\delta z_{m.\cos\gamma}$ $\delta z_{m.\cos\theta}$	Cosine error, measurement	X measurement error produced by misalignment of the X axes about the Z axis. X measurement error produced by misalignment of the X axes about the Y axis. Y measurement error produced by misalignment of the Y axes about the X axis. Y measurement error produced by misalignment of the Y axes about the Z axis. Z measurement error produced by misalignment of the Z axes about the X axis. Z measurement error produced by misalignment of the Z axes about the Y axis.	nm	12, 98
$\delta x_{m.dead}$	Dead band, measurement	A region of *true positions* over which there is no change in *measured position*.	nm	22
$\delta x_{m.drift}$	Drift, measurement	A change in mean *measured position* on a time scale of minutes or more with the *true position* constant. It includes the effect of *measurement temperature coefficient* and other environmental effects.	nm	17
$\delta x_{m.hyst}$	Hysteresis, measurement	The difference between *measured positions* when a single *true position* is approached from both ends of the range.	%	16
$\delta x_{m.lin}$	Linearity error, measurement	*measurement mapping error* with a first order power series (straight line) *measurement mapping function*.	%	8
$\delta x_{m.map}$	Mapping error, measurement	One half of the maximum peak to peak of *the residual curve* to the *measurement mapping function*, expressed as a percentage of the full measurement range.	%	9
$\delta x_{m.n}$	Noise, measurement	The 1σ spread of *position measurements* when the position being measured is stationary. It is a combination of *sensor noise* and *quantisation noise*.	nm	13
$\delta x_{m.ndens}$	Noise density, measurement	The RMS *Gaussian measurement noise* per unit *bandwidth*.	pm·Hz$^{-1/2}$	15
$\delta x_{m.ng}$	Noise, Gaussian, measurement	Contribution to *sensor noise*, generally with uniform *density* over the *bandwidth*, caused by thermal effects in electronic components.	nm	14
$\delta x_{m.nquant}$	Noise, quantisation, measurement	Contribution to *measurement noise* due to the finite number of bits used in digitisation	nm	16
$\delta x_{m.nsens}$	Noise, sensor	Contribution to *measurement noise*. A combination of *Gaussian noise* and *interference*	nm	13
$\delta x_{m.res}$	Resolution, measurement	The smallest measured *displacement* that can be resolved. This will be the greater of *measurement noise* or *quantisation noise*.	pm	13
$\delta x_{m.rpd}$	Reproducibility, measurement	The 1σ spread of the *measured positions* when the same true half range *displacement* is applied to the sensor repeatedly, approaching from both directions. Reproducibility includes *hysteresis*, *dead band*, *drift* and *repeatability*.	nm	16

Symbol	Term	Description	Units	Page(s)
$\delta x_{m.rpt}$	Repeatability, measurement	The 1σ spread of the *measured positions* when the same true half range *displacement* is applied to the sensor repeatedly, under the same operating conditions and in the same direction. Repeatability does not include *hysteresis* or *drift*.	nm	16
δx_{mA}	Accuracy, measurement	Arithmetic sum of *measurement trueness* and *precision*, e.g. $\delta x_{mA}=\delta x_{mS}+\delta x_{mR}$	nm	7
δx_{mpR}	Precision, measured position	The 1σ spread of *measured positions* including *measurement precision* and *position precision*	nm	18
$\delta x_{m\theta.abbe}$ $\delta x_{m\phi.abbe}$ $\delta y_{m\gamma.abbe}$ $\delta y_{m\phi.abbe}$ $\delta y_{m\gamma.abbe}$ $\delta y_{m\theta.abbe}$	Abbe error, measurement	Measurement error along the X axis caused by a rotation about the Y axis coupled with a Z *Abbe offset*. Measurement error along the X axis caused by a rotation about the Z axis coupled with a Y *Abbe offset*. Measurement error along the Y axis caused by a rotation about the X axis coupled with a Z *Abbe offset*. Measurement error along the Y axis caused by a rotation about the Z axis coupled with a X *Abbe offset*. Measurement error along the Z axis caused by a rotation about the X axis coupled with a Y *Abbe offset*. Measurement error along the Z axis caused by a rotation about the Y axis coupled with a X *Abbe offset*.	nm	12, 13, 24, 98
δx_{mR}	Precision, measurement	The 1σ spread of *measured positions* including *measurement mapping error, resolution, noise*, and *reproducibility* with the *true position* held constant.	nm	7
δ_{xmS}	Trueness, measurement	The difference between the mean *measured position* and the *true position*, $x_m\text{-}x_p$	nm	7
δx_{my} δx_{mz} δy_{mx} δy_{mz} δz_{mx} δz_{my}	Crosstalk, measurement	X measurement caused by a displacement along Y X measurement caused by a displacement along Z Y measurement caused by a displacement along X Y measurement caused by a displacement along Z Z measurement caused by a displacement along X Z measurement caused by a displacement along Y	nm	
$\delta x_{p.\cos\phi}$ $\delta x_{p.\cos\theta}$ $\delta y_{p.\cos\gamma}$ $\delta y_{p.\cos\phi}$ $\delta z_{p.\cos\gamma}$ $\delta z_{p.\cos\theta}$	Cosine error, position	A position error produced when the motion axis is misaligned with respect to the true position measuring device axis. X position error produced by misalignment of the X axes about the Z axis. X position error produced by misalignment of the X axes about the Y axis. Y position error produced by misalignment of the Y axes about the X axis. Y position error produced by misalignment of the Y axes about the Z axis. Z position error produced by misalignment of the Z axes about the X axis. Z position error produced by misalignment of the Z axes about the Y axis.	nm	12, 98
$\delta x_{p.dead}$	Dead band, position	A region of *desired positions* over which there is no change in *true position*.	nm	22
$\delta x_{p.drift}$	Drift, position	A change in mean *true position* on a time scale of minutes or more with the *commanded position* constant. It includes the effect of *position temperature coefficient* and other environmental effects.	nm	17
$\delta x_{p.hyst}$	Hysteresis, position	The difference between *true positions* when a single *desired position* is approached from both ends of the range. The maximum value is quoted, as a percentage of the range.	%	22-24, 75,76,82
$\delta x_{p.lin}$	Linearity error, position	*Position mapping error* with a first order power series (straight line) *position mapping function*.	%	19

Symbol	Term	Description	Units	Page(s)
$\delta x_{p.map}$	Mapping error, position	One half of the maximum peak to peak of the *residual curve* to the *position mapping function*, expressed as a percentage of the full positioning range.	%	19
$d_{xp.max}$	Range, measurement	The maximum *displacement* that can be measured.	µm	
$d_{xp.max}$	Range, position	The maximum *displacement* that can be generated.	µm	66,82,86
$\delta x_{p.n}$	Noise, position	The 1σ spread of *true positions* when the position command is stationary.	nm	20
$\delta x_{p.ndens}$	Noise density, position	The RMS *Gaussian position noise* per unit *bandwidth*	pm·Hz$^{-1/2}$	20
$\delta x_{p.ndrive}$	Noise, piezo drive	Contribution to position noise from the piezo drive electronics	nm	21
$\delta x_{p.ng}$	Noise, Gaussian, position	Contribution to *position noise*, generally with uniform *density* over the *bandwidth*, caused by thermal effects in electronic components.	nm	14
$\delta x_{p.nquant}$	Noise, quantisation, position	Contribution to *position noise* due to the finite number of bits used in digitisation	nm	26
$\delta x_{p.res}$	Resolution, position	The smallest *displacement* that can be commanded. This will be the greater of *position noise* or *quantisation noise*.	pm	15, 20
$\delta x_{p.rpd}$	Reproducibility, position	The 1σ spread of the *true positions* when the same true half range *displacement* is commanded repeatedly, approaching from both directions. Reproducibility includes *hysteresis, dead band, drift* and *repeatability*.	nm	22
$\delta x_{p.rpt}$	Repeatability, position	The 1σ spread of the *true positions* when the same true half range *displacement* is commanded repeatedly, under the same operating conditions and in the same direction. Repeatability does not include *hysteresis* or *drift*.	nm	24
δx_{pA}	Accuracy, position	Arithmetic sum of *position trueness* and *precision*, e.g. $\delta x_{pA} = \delta x_{pS} + \delta x_{pR}$	nm	17
$\delta x_{p\theta.abbe}$ $\delta x_{p\phi.abbe}$ $\delta y_{p\gamma.abbe}$ $\delta y_{p\phi.abbe}$ $\delta y_{p\gamma.abbe}$ $\delta y_{p\theta.abbe}$	Abbe error, position	Position error along the *X* axis caused by a rotation about the *Y* axis coupled with a *Z Abbe offset*. Position error along the *X* axis caused by a rotation about the *Z* axis coupled with a *Y Abbe offset*. Position error along the *Y* axis caused by a rotation about the *X* axis coupled with a *Z Abbe offset*. Position error along the *Y* axis caused by a rotation about the *Z* axis coupled with a *X Abbe offset*. Position error along the *Z* axis caused by a rotation about the *X* axis coupled with a *Y Abbe offset* Position error along the *Z* axis caused by a rotation about the *Y* axis coupled with a *X Abbe offset*.	nm	12, 13, 24, 98
δx_{pR}	Precision, position	The 1σ spread of *true positions* including *position mapping error, resolution, noise,* and *reproducibility* with the *commanded position* held constant.	nm	18
δ_{xpS}	Trueness, position	The difference between the *true position* and the *desired position*. In a *closed loop* system this will be the same as the *measurement trueness*.	nm	18
δx_{py} δx_{pz} δy_{px} δy_{pz} δz_{px} δz_{py}	Crosstalk, position	Displacement along *X* axis caused by a displacement along *Y* Displacement along *X* axis caused by a displacement along *Z* Displacement along *Y* axis caused by a displacement along *X* Displacement along *Y* axis caused by a displacement along *Z* Displacement along *Z* axis caused by a displacement along *X* Displacement along *Z* axis caused by a displacement along *Y*		
$f_{0\gamma}$	Resonant frequency, rotational	The frequency of the fundamental rotational oscillation modes around the measurement axes.	Hz	34, 89

Symbol	Term	Description	Units	Page(s)
f_{0x}	Resonant frequency, lateral	The frequency of the fundamental lateral oscillation modes along the measurement axes.	Hz	34, 89
$k_{xm.ndens}$	Noise coefficient, measurement	A system constant that enables the *measurement noise density* to be determined knowing the capacitor geometry.	$Hz^{-1/2}$	65,66
$k_{xp.ndens}$	Noise coefficient, position	A system constant that enables the *position noise density* to be determined knowing the capacitor geometry.	$Hz^{-1/2}$	65,66
$SF_{x.sys}$	Scale factor, system	A multiplier used to convert system units [volts in an analog system or numbers (*system units*: SU) in a digital system] to measurement units (μm)	μm$\cdot$V^{-1} or μm$\cdot$SU^{-1}	
SU	System units	The numbers used internally in a digital control system.	-	
$t_{xs.l}$	Settling time, large signal	Time taken for the *true position* to get to within 1% of its final value and stay there after a command step of 80% of the full range.	ms	41, 91
$t_{xs.s}$	Settling time, small signal	Time taken for the *true position* to get to within 1% of its final value and stay there after a command step of 0.5μm.	ms	37, 41, 90
$u_{xp.max}$	Slew rate	The highest rate of change of *true position* with time that can be achieved.	μm$\cdot$ms^{-1}	38, 40, 79, 92
x,y,z	Position	A point in space measured with respect to the *axes*	μm	5
X,Y,Z γ,θ,ϕ	Axes	A reference frame used for distance, motion and rotation measurement. Queensgate uses a right-handed orthogonal Cartesian system. The origin of the reference frame is the *true position* when the *position commands* x_c,y_c,z_c are zero.	-	5,17
x_c,y_c,z_c	Position, command	The *position* set-point sent to a controller	μm	18
x_d,y_d,z_d	Position, desired	The *position* that the user requires	μm	20
x_m,y_m,z_m	Position, measured	A mean *position* as measured by the system under discussion	μm	7
$x_m=A_XX_P$	Mapping polynomial, measurement	A power series representation of *a measurement mapping function*. The 'A' terms are row vectors *of measurement mapping coefficients* and X_A, Y_A and Z_A are column vectors of powers of the *true position* x_a,y_a,z_a.	-	8
x_p,y_p,z_p	Position, true	A *position* as measured with a perfect measurement system	μm	7,18,20
$x_p=B_XX_C$	Mapping polynomial, position	A power series representation of *a position mapping function*. The 'B' terms are row vectors *of position mapping coefficients* and X_C, Y_C and Z_C are column vectors of powers of the *command position* x_c,y_c,z_c.	-	20